PELICAN BOOKS

Genetic Engineering for Almost Everybody

William Bains was born in 1955 and educated at Mill Hill School, London, and Corpus Christi College, Oxford, where he obtained a first class honours degree in biochemistry in 1979. He took his Ph.D. in molecular genetics at Warwick University in 1982 and was subsequently awarded a Fellowship funded by the Muscular Dystrophy Association to study aspects of the molecular genetics of muscle development at Stanford University, California. He is currently a lecturer in biochemistry at the University of Bath and is researching further into molecular genetics, funded by the Muscular Dystrophy Group.

William Bains has red hair, short sight, and lives in Bath with his wife and baby daughter.

William Bains

Genetic Engineering for Almost Everybody

PENGUIN BOOKS

Penguin Books Ltd, 27 Wrights Lane, London W8 5TZ (Publishing and Editorial)
and Harmondsworth, Middlesex, England (Distribution and Warehouse)
Viking Penguin Inc., 40 West 23rd Street, New York, New York 10010, USA
Penguin Books Australia Ltd, Ringwood, Victoria, Australia
Penguin Books Canada Ltd, 2801 John Street, Markham, Ontario, Canada L3R 1B4
Penguin Books (NZ) Ltd, 182–190 Wairau Road, Auckland 10, New Zealand

First published 1987

Typeset, printed and bound in Great Britain by
Hazell, Watson & Viney Limited,
Member of the BPCC Group,
Aylesbury, Bucks
Typeset in 9½/11½pt Linotron 202 Melior

To my father, who asked me to explain what I was doing

Contents

Acknowledgements

Although the foibles, errors and *non sequiturs* of this book are entirely my own, they would have been far more numerous without the generous help of Paul Barnett, Dr Robert Old, William Kaufmann, several (wisely) anonymous reviewers and the good folk at Penguin. My thanks also to my wife Jane for putting up with (among many other things) my nocturnal t-t-typing.

Chapter 1
Introduction

In the mid 1960s a growing number of scientists said that we were on the verge of a computer revolution. The technology of building and programming computers had become so advanced that we could look forward to the day when computers would be an integral part of everyone's life. Why, computers for use in small businesses, even in private homes, could be a reality before the end of the century.

In the mid 1970s another growing band of enthusiasts said that we were on the verge of a biological revolution. The technology of 'genetic engineering' had become so advanced that we could look forward to the time when we could build organisms as easily as we build computers today; the new 'miracle drugs' and genetically engineered vaccines would make such changes in medicine that by the end of the century the only major illnesses left would be heart disease and old age.

Here we are in the mid 1980s. Computers are among us, but so is cancer. Every household appliance has a 'chip', but we still have influenza.

What happened?

This book is about what happened, and why. The ideas are simple at root, and, unlike the story of the computer revolution, we will find ourselves continually being drawn back to the real subject of biology — human beings and their companions on the Earth. And it is because of this, because genetic engineering is about life and not about abstract mathematics, that the biological revolution has proved so baffling to professional soothsayers.

1

The difference between a computer and a biological revolution is the difference between a robot and a man.

Make no mistake; the genetic engineering revolution *has* arrived. It began in the period from 1969 to 1971 in two academic research laboratories, where Paul Berg (at Stanford, California), and Stanley Cohen and Paul Boyer (at Stanford and the University of California at San Francisco (UCSF) respectively) were thinking of applying new biochemical techniques to some long-standing problems in biology. They wanted to study the fundamental mechanisms of the working of genes, problems pondered by thinkers since before the birth of Christ. Why are children similar to their parents, but not identical? Why do some viruses cause severe disease, and others no disease at all? Why do we grow with such marvellous precision from a tiny fetus to an adult, and then fall prey to old age? The new techniques began as better ways of trying to answer some of these questions, and as such they were of little use to the world outside the universities. Now, if they produced some *answers* . . .

They *did* produce some answers, although these large problems are far from being solved. But the new techniques also produced something else, because they were different from any which biochemists had used before. They were not only a way of attacking a specific academic problem but also a set of general methods of manipulating genes – they were the tools of genetic engineering. Very early in their development some scientists realized that they had invented something of more than ordinary importance, and possibly of some considerable danger. In 1973 two crucial meetings were held in which scientists involved in the new technology discussed, often with some heat, whether this development should go ahead at all. The second meeting, at the Asilomar Conference Center on the central Californian coast, included lawyers, philosophers and the press, an extraordinary gathering for a technical discussion. Two decades earlier the science that had generated the technology of genetic engineering, a subject called 'molecular biology', had not even existed, and until 1973 it remained the province of a small band of devotees who had produced a lot of experimental results which they found very interesting but which were of no actual use. All that changed with Asilomar.

That is probably when you first heard the phrase 'genetic engineering'. We were all treated to an explosion of media interest in a subject that some commentators found hard to spell, let alone understand. 'Recombinant DNA', 'plasmids' and 'clones' were pulled from the dry prose of scientific journals and turned into instant headlines. The presenters of numerous science programmes and newspaper columns told an astonished world that it would never go hungry again, that within a decade cancer would be a thing of the past, or that the new technology threatened the stability, or even the existence, of life on Earth. At hearings in Cambridge, Massachusetts, on the safety of recombinant DNA techniques, groups of protesters holding banners reading ' "We will create a master race" – Adolf Hitler' tried to shout down speakers who claimed that recombinant DNA heralded the dawn of the Age of Aquarius, the millennium of our time.

What was it all about? Were these claims supported by facts, or by the scientists who were producing the new technology? Certainly a few of the scientists went nearly as far as the press. Edwin Chargaff, one of the founders of molecular biology as a subject, claimed that recombinant DNA was violating the accumulated evolutionary wisdom of a billion years, and that we meddled with such things at our peril. Paul Berg, a developer of the new techniques, said that these were mere metaphysical considerations and in 1974 added, 'I'd stop if there was a sound practical reason, but not if it were an ethical judgement.' In such a new science, there *were* no 'sound practical reasons', as there were no facts, no history to go on. There was the normal caution of science to temper judgements, but too often that caution was thrown to the winds.

After the initial euphoria, it was inevitable that a let-down would follow. Somehow a potential cancer cure can remain potential for only so long before someone starts asking who is fooling whom. Similarly, the hypothetical catastrophes remained hypothetical, defying the doom-watchers by refusing to wipe us all out. Of course, it takes many years for new biochemical processes to become commercial realities, and the fifteen years' testing that an average drug requires before it is released for public use is far longer than is needed for a new computer to appear on the shop counter. But even given this constraint, progress seemed unpromisingly slow. The new wonder drugs had not even appeared in the laboratory, let alone left it for clinical trials. By 1980 only two of the dozens of miracles we had been promised were still visibly progressing. In fact the let-down was the result of unrealistic expectations being exposed to the cold light of scientific fact, although the cause was unimportant to the emotions of investors who had stampeded to buy stock in genetic engineering companies and to the pundits who had said that this was the start of civilization as we would come to know it. Genetic engineering was no longer trendy – it appeared to be going the way of other Californian enthusiasms like Haight-Ashbury and Flower Power.

This was unfair. In 1985 more applications of genetic engineering were presented to the world than during the entire period of the 1970s enthusiasm. Useful (though not miraculous) clinical products began hospital testing; several finished such testing and began the build-up to marketing; one was released for sale. This is quick work for the biomedical industry. Tremendous advances had been made towards the application of genetic engineering to a range of problems, like diagnosing disease and improving crop plants. But the glamour had faded. With an alarming failure rate in the new golden industry of tomorrow, with research companies falling on hard times – even going bankrupt – and with stock falling, it seemed that the age of miracles had again failed to arrive.

Then in 1985 it all started up again. Donald Wood of the Muscular Dystrophy Association announced in 1986 that new breakthroughs in recombinant DNA techniques would mean that a detection system for carriers of muscular dystrophy would be available 'in a few months to a year'. Other diseases were on the brink of a foolproof detection system, and scientists were again talking about the application of genetic engineering to the breeding of sheep and pigs, and conducting 'gene therapy' on humans. Enthusiasm surged once more, the old headlines of doom or of Utopia were dusted off and pasted into place, and genetic engineering was again the wonder of our age.

Is this new euphoria justified? Was the old euphoria justified? This book is aimed at finding out. We will approach the new technology in the same way as the scientists did, and come to some sensible conclusions about what it can and cannot do. These conclusions will be – must be – based on facts, not on speculation. If this means that we will have to say, 'We do not know', then so be it. This, then, is a technical appraisal for non-technical people. We shall look at the problems we wish to solve, the methods available for solving them, and the results achieved so far. There *have* been results, pouring out of the laboratories at an increasing rate, and there *have* been medical breakthroughs which will lead to new technologies for fighting disease. But much of the hype was not justified then and it is not justified now. We can say that with some confidence, now that the technology of genetic engineering has been in use for a few years and some of its strengths and weaknesses have been revealed. Of course, I can no more predict the future of biotechnology fifty years from now than can anyone else. The biotechnology revolution is one of the fastest moving and most exciting areas of science today, and who knows where it will be in fifty years? But, unlike the scientists of fifteen years ago and very much unlike the headline-writers of ten years ago, we now have a fairly good idea of where the science will and will not be going over the next two to five years, and where it might head after that. We will look at the facts and then draw some conclusions. The facts begin with the science called 'molecular biology'.

Chapter 2
The Message in the Molecule

Molecular biology — what a curious title! We know what biology is about. It is about breeding rabbits or collecting frogspawn or studying wildflowers. And we know what molecules are — invisibly small particles of matter whose esoteric properties are the basis of chemistry. So how can I present a subject called 'molecular biology' to you? Even stranger is 'genetic engineering', which tries to impose a vision of hammers and cranes on the problem of why Granny had red hair.

Nevertheless, both phrases are good descriptions of an approach to the study of life which has been increasingly useful in the twentieth century. That is the molecular view of life. 'Genetic engineering' is the popular name for one of the practical applications of molecular biology. Although now a branch of the life science, molecular biology has roots, going back a century or more, in chemistry. For it does not treat life as a subject so complex that it can only be observed and catalogued but never explained. Instead, molecular biology assumes that life, at base, is the end result of the properties of the materials from which it is made, and that we can understand it by treating it as a collection of complicated molecules acting together rather like miniature clockwork.

Life, it says, is a molecular machine.

Now, when you meet someone in the street you do not see a vast tangle of tiny gears and levers whirling round like microscopic Meccano to pull the face into a smile. But if we ask why those muscles move as they do, why the eye sees what it does, then one approach is to look not at the whole person but at his or her components – the muscles, the nerves – and deeper into the parts that make up that person. In the same way, we drive a car as a single machine, but in order to repair it we mentally divide it up into suspension, electrics, cooling system and so on. And the ultimate constituent parts of the human animal are molecules.

Of course we can go overboard and say that a person is only a heap of molecules and is no more 'really' alive than any other heap of molecules, like a corpse or a brick. This is foolish; it takes a perfectly good way of looking at the world and uses it out of context. We have many different ways of looking at people in everyday life – what would seem reasonable behaviour for a politician on TV seems bizarre for Granny to do in the living-room. We can watch hideous dismemberings as part of a film, but feel queasy if a loved-one cuts their finger. Scientists have found that, to discover how genes work, or on occasion fail to work, looking at people from a molecule's-eye view is a very productive exercise.

For the studies described in this book, the molecule's-eye view has turned out to be rather useful. Consequently, we will start from this viewpoint and watch the vision of life that it reveals unfolding until we reach its practical conclusion. We will not be discussing every detail of molecular biology: indeed, there are a myriad details which make the subject one of amazing complexity, an array of patterns and functions which science is only beginning to unravel. Rather, we will be painting a broad picture of the patterns and styles in the molecular machine that is life, a picture that is surprisingly simple; for there are a few great themes in that machine, and we need only look at these to follow them to their practical end.

We will start at the very base of the molecular machine, at the genes. Genes are the ultimate arbiter of what we are, and are the rulers of our molecular Meccano. (Which came first, the chicken or the egg? Neither – the chicken's genes came first, and everything else followed from that.) Of course, other things influence our lives – we are not simply genes. However, the genes lie at the base of all we are, and so have been of great interest to science. The study of genes is the science of genetics, and it is one of the most misrepresented of the life sciences. Many people 'know' that it is terribly complicated, or that it is all about evolution, or that it may be something to do with eugenics. Like other academic subjects, genetics has accumulated a jargon which enables its practitioners to talk to each other precisely, but which forms a dense jungle of terms which can take the student years to penetrate and which most people do not have time to master. The deeper aspects of genetics can indeed be terribly complicated but, as I promised, the basic idea behind it all is so simple that you may wonder why it needs academic study at all.

The basis of genetics is this. Parents pass some of their distinguishing features on to their children.

This fact has been obvious for millennia. The art of breeding racehorses or flowers relies on it. If a steeplechase champion covers a flat-race mare, we are confident that the foals will be pretty fast. A farmer crossing two high-yielding types of potato can reasonably expect a plant that yields a good crop. This rule-of-thumb method had been elevated to a fine art long before science was born, producing the multitude of 'races' of dogs and garden flowers we see today. In the course of two thousand years plant-growers used this technique to turn the pineapple from a tiny, bitter-fruiting body into the huge, sugar-laden product we know today. Maize is a plant so carefully selected by Central American farmers that it no longer has an exact counterpart in the wild, and has to be artificially propagated to survive. We created this new strain simply by selecting the plants which gave the largest cobs and the best flour. The farmers of the New World knew that if they bred from those prize plants, their special characteristics would be inherited by their descendants.

In this century breeding has had much scientific use, often employing more subtle techniques than those available to pre-scientific farmers – techniques developed by the science of genetics. Farm animals have been tailored for higher meat, milk or wool production according to complex schedules of cross-breeding, usually mediated by artificial insemination rather than slower, more chancy traditional methods. While traditional breeding took, say, seven thousand years to turn a small, wiry breed of forest boar into the domestic pig, modern genetics can improve the pork yield of those same pigs by anything from 10 to 50 per cent in a few decades. Curiously, modern breeding methods often result in animals or plants which do not themselves have the characteristics farmers are looking for, but which are well suited to be the *parents* of the animal or plant with those particular qualities. Two quite separate lines of animal or plant are chosen by the breeder to produce the offspring that will actually be of use to the farmer. Cattle are often reared on such breeding plans, as are many grain crops: 97 per cent of the maize planted in the United States to harvest for corn is hybrid maize bred in this way. Indeed, many breeders consider that their art is so advanced that new techniques like genetic engineering will be of little use to them. Later on, we shall see if they are right.

The original breeding programs that led to modern potatoes and pigs worked so well because they had a vast number of genes to draw on for selection. The term 'gene' is used to describe the basic units of inheritance; in traditional genetics, one gene is considered to govern the inheritance of one characteristic. Nearly every aspect of our body is affected by the genes we carry. To illustrate, think how many genes there are in a human being (this is still a controversial point among the experts, but we can get a rough idea). How many characteristics did you inherit from your parents? The shape of your face? Yes, but is that one characteristic? No. The nose, the cheeks, the dome of the skull — each has its individual shape and clearly we can inherit each feature separately, as we can say of someone, 'He's got a nose just like Smith's, and ears rather like old Harry's.' Each of these independent shapes or forms is probably governed by several genes for bone size, shape, thickness and so on. Even the most general characteristics are also determined by genes. Your genes told the infant you once were to grow into a human being, not into a dog or an oak tree — that is why human beings only produce other human beings. We pass on a whole collection of 'basic humanity' genes to our offspring. There are other genes which determine how fast you grow, and to what size. We know all this because there exist variants of all these genes which do not quite work as they should, so the people concerned end up too short or too tall, or with the wrong number of fingers, or any one of a large range

9

of abnormalities. These genes must have been changed at one time from their correct version to their present incorrect version. This is known as 'mutating', and the changes produced are called 'mutations'. The abnormal characters caused by a mutation, a change producing a 'mutant gene', are passed down faithfully from parent to child (if they do not make the parent so ill that he or she has no children, which is sometimes the case – we will discuss this in Chapter 5), just as the normal genes are passed down by other parents to their children. There are tens of thousands of such known variants, and probably many more that have not been catalogued. As mutations are changes in the genes, this means that at the very least there are tens of thousands of genes in each of us.

All our characteristics, from the most basic features of our humanity to tiny details that made us individuals, are influenced to a greater or lesser degree by our genes.

Now this raises a problem. If your father is short and your mother is tall, and these two characteristics are determined partly by genes, then clearly you cannot be short *and* tall. Either you inherit *only* your father's 'height genes', or *only* your mother's, or there is a compromise. Nature does both. Each individual contains duplicates of nearly all their genes: the gene for short sight, for example, is present in two copies in every cell – except red blood cells – of a short-sighted individual. (I am talking here about people with severe short sight, the kind that requires children to wear glasses at the age of eight or ten and makes them incapable of seeing clearly beyond the end of their nose by the time they are twenty. Mild short sight, requiring glasses for driving but not reading, for example, is not inherited in the same way.) If we have children, we pass on just one of these copies to them. Thus they get one copy from one parent, one copy from the other, and so end up with two. There is therefore some compromise: we pass one of our two copies of each gene on to a child, so that the child inherits a combination of the genes from both parents. This means that if a person with two copies of the short-sight gene has a child by a partner who has two copies of its normal-sight counterpart, the child will inevitably end up with one normal-sight gene and one short-sight gene, one from each parent. Here the compromise breaks down, and a decision has to be made. In this case, the decision is always made in favour of the normal-sight copy so that the short-sight version invariably takes a back seat. In genetics, the short-sight gene is referred to as 'recessive' and its normal-sight counterpart as 'dominant' to designate their respective roles.

This is a very simple rule, although it was only worked out, by Gregor Mendel, in the middle of the last century. But it gets a bit more complicated if more than one gene is involved in the regulation of one characteristic. For example, how tall a person grows is determined by many features, each of which is influenced by at least one gene. So what decides how we inherit such complex characters? In this case, each gene copy is governed by the dominant/recessive rule, but, as the final result is an average of the effects of a whole lot of genes, the child's inheritance is the complex result of adding together a number of genes, each with a different dominant/recessive rule.

Whether a gene is dominant or recessive depends on the details of the working of the gene, and in many cases we do not know how they work. In the case of the recessive gene for short sight, it seems likely that this gene cannot do something (although no one knows what) which the dominant, normal gene can do. So if you have a normal and a short-sight gene, the former can cover for the latter's incompetence. But things are not always so simple. It might be thought that dwarfism was also caused by a lack of something (and indeed in Chapter 4 we shall mention that some types of dwarfism *are* caused by a lack of something). But the mutant gene which causes one sort of dwarfism (called achondroplastic dwarfism) is *dominant* to the normal gene. Therefore normal people may lack something which achondroplastic dwarfs possess. So there is no simple rule that the 'normal' gene is always dominant.

Thus, while the workings of one gene can be quite simple, the workings of a number acting in concert can be very complex. As most of our observable characteristics are the result of the action of several genes – things like height, skin colour, weight, head shape and so on – this makes genetics rather complicated. Only rarely do we see a characteristic which is under the control of only one gene: short sight is one example.

As I mentioned, we do not pass *all* our genes on to our children. We only pass on one copy of our own two. What if we pass on a copy of a recessive gene, and the other parent passes on a copy of the dominant partner of that gene? Then our children will show whatever characteristic is determined by the dominant gene, and not whatever characteristic our recessive gene produces. Our own gene will seem to have disappeared! However, it is still there as the 'silent partner'; it simply is not making itself evident at the moment. This is how genes can sometimes seem to skip a generation, an important point in medical genetics which we will consider in Chapter 5.

The pattern of inheritance outlined above applies to most of the genes we shall be discussing in this book. These include mutant genes, which can therefore be classified as dominant or recessive just like other genes. If a mutant gene is a dominant one, then any person inheriting a copy of that gene will show the effects of the mutation. The mutant genes which cause the disease achondroplastic dwarfism and Huntington's chorea (see page 62) are of this type. An afflicted parent has a 50:50 chance of passing the gene on to any of their children (because he or she has two copies, one mutant and one normal, and which one is passed on is up to chance), and so the children themselves have a 50:50 chance of developing the disease. However, if the mutant gene is recessive (as is true of the gene for cystic fibrosis – see page 61), then the child needs to inherit the mutant gene from *both* parents before developing symptoms. If it inherits only one copy, then the normal copy of the gene will override the effects of the mutant copy and the child will appear normal. This has a good side – the child will not suffer this cruel disease – but it has a bad side, too. The child is now a *carrier* of the disease, because it is carrying one copy of the mutant gene, and each of its children will have a 50 per cent chance of inheriting the gene. Thus they still have the possibility of having the disease cystic fibrosis, even though their parent did not have it. In many cases the carriers cannot be distinguished from people with no recessive mutant gene at all, so the diseases these mutations cause can crop up unexpectedly in families with no known history of the illness. While this is not very worrying if the recessive gene concerned is for short sight, it is tragic if the gene is the cause of cystic fibrosis. We will be coming back to this problem of detecting carriers of genetic disease again later, as it is an area where genetic engineering has been able to play a major role.

Not everything is determined by our genes, of course. The general colour of your skin is genetically determined, but a few months on the beach in Miami could alter that colour quite a lot (unless you are already reading this in Miami, of course). Indeed, many characteristics may be influenced by the environment as well as by genes. Short sight is an example of this. The short-sight gene is actually a gene which makes its possessors very much more prone to develop a particular type of severe short sight. But altering environmental influences – the amount of time the eyes are used for close work during adolescence, the type of corrective lenses used and so on – can markedly influence the course of development of the myopia. Indeed, in a sense environment can overrule the genes entirely, for what are glasses but a feature of the short-sighted person's personal environment, one that renders them normal-sighted? A more contentious point is how much our intelligence is due to the genes we carry and how much to the environment we live in. The short and scientifically correct answer is not only that we do not know, but also that the question is probably meaningless. Both genes and environment play a combined role in making an individual's unique mental capacities, and they cannot be separated any more than the yolk and the white of an egg can be separated in an omelette. However, we do know that genes play some role, as otherwise animals with completely different genes – like dogs or hamsters or beetles – would have the same ultimate capacity for intelligence as we do, which they do not, and there would be no genetic diseases which cause mental retardation, which, tragically, there are.

So far we have talked about genes in terms of their effects. A gene 'causes' short sight, or 'makes' you tall, but that does not really tell you anything about how the gene works. After all, genes were originally defined simply as those things which make you have the observed characters you do have, so in effect all we have said is that whatever causes you to have short sight acts by causing you to have short sight. It is very useful to know that such a thing as a gene exists, and that variation between members of a species is not due to chance or to some averaging-out of the characters of the parents. And we can determine the *rules* governing how genes operate without having the least idea how genes work, just as we could work out the traffic rules in New York without ever knowing how drivers are trained. However, we have seen that genes are pretty central to everything that we are, and to everything that every other living thing is. So if we are to step beyond description we must try to find out how these things called genes operate.

Again, we will use the gene for short sight as an example. It acts as an on/off switch – either you have the gene, or you do not. We can get an idea of what is being switched by considering what short sight is. After all, eyes are not made of genes, any more than the glasses we put in front of them. Instead they are made of what the rest of us is made of (we will come to what that is in the next chapter), but organized in a particular way so that an eye, and not a fingernail or a flower, results. The short-sighted eye is not organized quite correctly, however, as somewhere in its growth it did not follow the usual plans, ending up a bit more egg-shaped than the normal eye and so unable to focus properly. Thus the short-sight gene is providing the organization needed to make a faulty eye, or, more sensibly, is providing a faulty version of the correct organization needed to make a normal eye. In short, the gene is related to *information*. The 'normal eye' gene ensures that the eye develops correctly by providing a piece of information about the shape in which the eye should grow. When the body has one of these normal genes to use, it employs it as part of the 'plans' for an eye; but when it only has short-sight genes to go on, it has to use the faulty plans they provide instead. Thus the short-sight gene is simply the normal-sight gene with a mistake somewhere in it, a change, a mutation. Once, no doubt, many hundreds of years ago, this particular gene was a perfectly normal gene, but then the mutation occurred, changing it to a faulty version which has been passed down ever since.

It took a physicist, Erwin Schroedinger, to point out that all genes provide information: biologists had used this as a rule of thumb since it was proposed by August Weissman in 1893, but he elevated it to a general principle, saying that whatever the physical basis of the gene was, its most basic function was to provide information.

The amount of information needed is enormous. Consider how easy it is to turn a pig into sausage. All you need is a mincer, a press and some skins – and these could be described on a sheet of paper. But what would a machine for turning sausage into a pig look like? A car assembly plant would be a child's toy in comparison. Whole factories would have to be devoted to making one ear. Yet such a machine exists: it is the pig itself. The detailed plans needed to build and run an enterprise to turn sausage (or any other raw material in the pig's varied potential diet) into living pig are contained in the genes of the pig. No wonder there are so many genes in a pig, or in a human. The task they must coordinate is a monumental one. Genes are not just a set of plans for making short-sighted eyes, but a pile of plans kilometres high for making an entire living being.

Somewhere in the pile of plans that describe how to make a man is a page with the heading: 'How to make eyes, #8376. Focal length.' In the short-sight gene this page has a misprint some-where on it, so that the focusing does not come out quite right. Somewhere else there is a page on 'How to make hair, #8222. Making RED hair.' When the growing child gets to the hair-making stage, it will use this page and not any one of the several other variants which can direct it to make black or brown or blond hair. As the genes are the ultimate databank of life, the result of this set of genes will inevitably be a person with red hair, unless, of course, that person alters their environment suf-ficiently to change it – by using dye or bleach.

Of course, the genes do not *literally* say that. We will discuss what the genes do actually say in the next chapter. However, we can consider a particular gene as reading: 'How to make eyes, #8376. Focal length', because it directs some action to be per-formed which, if it were not carried out as described by the gene, would result in an aberration of the focal length of the eye but not of, say, the shape of the stomach or the colour of the skin.

Genes must be organized. A tonne of blueprints is of no use to us unless we can find the one we want. So the science of genetics studies not only what genes are, but also the way in which they are organized into files and boxes: the record-keeping of the body. The analogy between genes and files is close. Both are repositories of data, information on possible courses of action. In this electronic age, files need not be sheafs of paper in a metal cabinet. Cards, paper, or magnetic tape and disks can all hold information arranged in 'files', and the same information may be transferred from one kind of record to another. This book has been held on magnetic disk in a word-processor, as scribbled words on loose sheets of paper, as symbols in the printer's computer and as the bound set of pages you are reading, but it remained the same book in all these incarnations. If the information is identical, no matter how you store it, is there any way of finding out what the information itself looks like when it is separated from its storage medium?

Can we not separate the essence of a plan from its method of storage, the message from the medium?

We can't, of course. Information is not a 'thing', like an egg or the Taj Mahal. It is something that is held by a thing: it is the way in which a physical object is organized. If we take a page of print from a book and remove the physical manifestations, the paper and the ink, we are not left with 'pure information', only with a hole in the book. This may seem pretty obvious, but so far we have been treating genes as if they were 'pure information'. What Gregor Mendel called the gene, by observing its effects, is a piece of information, which must be held on some corresponding piece of matter. The genetic file must be written on genetic material.

The genetic material, the physical basis for the information we call 'genes', is DNA.

DNA (for deoxyribonucleic acid) is a chemical present in all living organisms. (Some viruses do not have it, but then viruses are not really living.) The genetic material was once thought to be beyond the realm of chemistry because of the almost magical attributes it would have to have, but now it is understood to be a molecule whose quite extraordinary properties derive from ordinary chemical laws. The discovery that DNA was the genetic material was made by Oswald Avery in 1944, and in 1953 James Watson and Francis Crick worked out what the molecule of DNA looked like and how its structure might lead to its peculiar properties.

The DNA molecule is vast by molecular standards – a huge number of small units linked into a long molecule, resembling links in a chain. Each whole DNA molecule consists of two such chains wrapped round each other like strands in a rope, a 'double helix'. The basic units of these chains are called 'bases' (more correctly they are called deoxyribonucleotides, but 'base' is the universal nickname). There are four different types of base in DNA, called adenine (A), guanine (G), thymidine (T) and cytosine (C), and by altering the order in which these four are strung together we can alter the information stored by the DNA. Thus A, C, G and T act as letters in a four-letter genetic alphabet. The language of the genes is a language which can convey messages of great subtlety to our bodies in 'words' containing only these four letters.

Apart from its use of four letters instead of twenty-six, the language of DNA contains features found in many more mundane languages. There are special codes which act as punctuation marks, saying 'Stop' or 'Start here', and others which act as brackets, to indicate that 'The next few words are not essential, skip them' (like this). There are special ways of ordering genes which perform related functions into closely packed sections of DNA, like paragraphs or chapters. Indeed, the scientists who elucidated all this came to regard the study of DNA not as part of chemistry but as an exercise in cryptography, the cracking of the 'genetic code'. In this case the code consisted of only four letters. They only had to crack the code, and the meaning of the genes would immediately become clear. Although the language of the genes has turned out to have rather more surprises than that, the way in which DNA is used to make the other major type of molecule with which we are concerned in this book, the proteins, is still called the 'genetic code'. So genes can be thought of as being like a written description of building a ship or mowing the lawn, a series of instructions written in a code of a few

17

simple symbols whose order can carry an infinite number of meanings, inscribed on a long molecule like ticker-tape or magnetic cassette tape.

It is a pretty big cassette.

The actual DNA molecule is only a quarter of a millionth of a centimetre across and each base is a thirtieth of a millionth of a centimetre long. Even so, if all the DNA in a fertilized human egg could be joined into one long double helix it would be two metres long. If this were magnified until it appeared as wide as a cassette tape, it would be over 2,800 kilometres long and would take more than two years to play on a tape recorder.

Every child will need its own DNA, and so its parents will need to provide it with a copy. For a machine, the task of copying such a tape would be an almost impossible one. Remember that the message in this 'tape' of DNA consists of the genes themselves. Any mistake in them is a mutation, and many mutations are extremely dangerous to their carriers. How can we arrange for the accurate copying of 2,800 kilometres of cassette tape?

Well, of course, it is not done without error. Occasionally changes are made by accident in the program of the genes – otherwise no mutations would ever arise. But not very many genes are affected by a mutation whenever the whole DNA 'set' is copied, so the copying mechanism must be very accurate. The way in which this is done is by use of the chemical properties of DNA. DNA is different from a cassette tape in that it comprises two interwound chains. Left together they look unremarkable, but if you unwind them the difference becomes apparent. The bases in DNA 'recognize' each other, and have a physical affinity for each other that results in their tending to stick together. They stick together in pairs, A with T and G with C. They do this because the two pairs have complementary shapes, like a lock and key, so that A only fits together with T and not with G or C: these pairs are called 'complementary base pairs'. In double helical DNA the bases are arranged facing each other in such a way that each base is always opposite its complementary base partner. Thus if one chain has an A at a particular point, the other chain will have a T opposite the A.

When the DNA chains are in a double helix, each base is nicely paired up with its complementary partner. But what happens if you separate the chains? The obvious thing to happen is that the chains will snap back together again, to get those bases properly paired up with each other. If you do not allow that to happen, however, then they will form pairs with bases which are not in DNA chains. Thus an A in the DNA chain will latch on to a T which is *not* in the chain, but which is just floating around as an isolated base. When all the bases in the DNA chain have done that, then they are all paired up again with their correct complementary bases; but those bases are not joined up. So we join them up. That results in a new DNA chain, with the bases in exactly the same order as they were in the old chain we took away! Thus we start out with a double helix of DNA, and 'unzip' it. The two single DNA chains latch on to their complementary bases, we join the bases up and we then have two identical DNA double helices. We now have a duplicate of our DNA molecule.

In practice this does not just happen by itself – it needs the help of an enzyme (which we will meet in the next chapter). And the whole DNA molecule is not duplicated at once, but a bit at a time. However, these are details of mechanism. The key to the process is that the bases of DNA can latch on to each other in complementary pairs, and so arrange a new DNA chain in the correct order without the need for any other information or help.

As information is stored in DNA by putting the bases in the right order, this method of self-duplication not only gives us two pieces of DNA, but also ensures that both pieces contain exactly the same message. Thus DNA is not just a very convenient way of storing a lot of information; it is also a very effective way of copying it. The self-duplication of DNA is crucial, as new DNA is needed whenever an organism is growing or reproducing. When you cut your finger, the damaged tissue needs its own copy of your DNA to tell it exactly how to repair the damage. When a child is conceived, all the tissues which will form its body need their own DNA to tell them where to go in the growing embryo, and what to do there. And, of course, the fertilized egg needs two sets of DNA to start off with. As we have seen, the new child gets one of these from each of its parents, and they can give slightly different versions of the information. Both these versions are passed out to all the tissues, where they are acted on according to the rules of dominance and recessiveness we have already discussed. We shall see in Chapter 5 how these principles (so far presented simply as rules, with no reason behind them) can in some cases be related to what is going on in the DNA itself.

Thus the existence of characters which are passed down from one generation to the next is possible only because those genes represent information – instructions coded on a molecule which can duplicate itself: the molecule of DNA.

When 'gene' was just a name for something we did not understand, we could only study DNA at a distance, by its effects, and could not hope directly to alter its functions. That is not to say that scientists knew nothing about genes – on the contrary, geneticists had a great store of information about the genes of many organisms, and they have used that information to improve farm stock enormously in the twentieth century. Mathematical studies of the frequency with which genes crop up in populations have provided a firm base both for the theory of evolution and for our expectations of what applied genetics can and cannot be expected to do. But all this was done by treating the gene as a 'black box', a mysterious something which had an effect but whose mechanism was unknown. Scientists could not point to a molecule and say, 'There is the gene for red hair', and consequently they could not manipulate that or any other gene directly, relying instead on indirect selection.

Altering genes directly is, of course, the science of genetic engineering.

In principle, what we have discussed in this chapter shows that such a direct alteration should be easy to make. A gene is simply a piece of information coded on a length of DNA. If we remove a section of DNA and replace it with another piece containing a different order of bases, we will have altered a gene. It would be the same as altering the music your tape deck is playing by changing the cassette. But, unlike a cassette, the gene will not have to be taken out and copied every time we want a copy for a friend's collection. For the DNA in which our 'new' gene is encoded will duplicate itself in just the same way as every other bit of DNA in that organism – indeed, it will be identical to every other piece of DNA there, as far as its chemical composition is concerned. So if we want to alter all the descendants of one experimental mouse, we need only alter the DNA in the 'germ cells', the sperm or the ova, of that first, ancestral mouse and the altered DNA we have made will duplicate itself with our alteration in it. The mouse will pass that altered DNA, that altered gene, on to its offspring (only to *half* its offsping on average, unless we manage to alter both copies of the gene in our original mouse). The same would be true if, instead of altering a gene that is already there, we added some more DNA to the mouse's existing complement of DNA. The information in that DNA would become part of the mouse's collection of genes, and could be passed on to its descendants like any other gene. Indeed, this latter experiment has been performed – with rather startling results, as we shall see in Chapter 11.

This is the fundamental difference between altering genes, which is the basis of genetic engineering, and surgery or chemistry. We could have dyed our mouse blue, but in doing so we would not have made a blue mouse any more than a man in a parka is a woolly man. We would just have produced a disguised mouse. But if we could alter the instructions in the mouse's haircolour genes from 'Make BROWN hair' to 'Make BLUE hair', not only would the mouse end up with blue hair, but its hair would *always* be blue, and the hair of some of its offspring too. We would have produced a race of blue-haired mice. We would have to dye our mouse brown if we wanted to get a brown mouse back.

Amusing though the idea of a Walt Disney-style blue mouse may be their production is not a project any scientist would undertake because it is pointless and probably impossible. But what useful, attainable research could our putative genetic engineer carry out instead, and how can we decide whether a genetic engineering project is easy, difficult or impossible? There must be limits – try as we might, we could not make a gene that could render a mouse invisible.

A second problem is really a more fundamental restatement of the first. How do we write instructions in the language of the genes? How do we order the four bases in the DNA chain to govern a mouse, or any other organism?

The answers to these questions are not simple, and they vary according to the problem we choose to attack. Indeed, the successes of genetic engineering in the 1980s have lain as much in finding problems to which there are simple solutions as in finding solutions to problems which look interesting. To see where these possibilities lie, we must look away from the gene and towards the mechanism it controls.

Chapter 3
The Clockwork Cell

The butcher's shop, with its meat on display, is a good place to start on an exploration of the animal body. A glance across the counter can distinguish hamburger from steak, although both may contain the same meat. The hamburger and the Sunday roast look different, whatever the shape of the former, because of their texture. A whole animal, or any sizeable chunk of it, is not homogenous like jelly or concrete, a solid statue of flesh, but is a mosaic of different types of material – the bone, the muscle, the skin.

This subdivision can be seen on a tiny scale. With a magnifying lens the first biologists saw that the cork of tree bark is divided up into tiny units with woody walls, like monks' cells, and they called these units 'cells'. More powerful microscopes showed that all living matter, not just tree bark, is divided into cells, which join together in huge numbers to form the tissues of the body like bricks in a wall.

These cells are the basic units of living matter.

During an operation, a human being may be kept alive even though parts are missing – the heart itself can be removed for over an hour and the patient remain clearly alive. So a whole body cannot be the basic, indivisible unit of life. The same is true of many tissues. Sections of bones, lengths of intestine, or nearly all the liver may be removed by a surgeon while the rest continues to function. Scientists can even separate individual cells out of tissues and keep them alive for months outside the body. They grow and multiply in the test-tube (usually, in fact, in plastic dishes of various shapes), and show just as much evidence of being alive as do the tiny creatures that swim around in pond water. So cells are alive in a real sense. However, the subdivision stops at the level of cells, for if part of a cell is chopped out the fragments remaining are not alive because they cannot perform the feat which is so characteristic of living cells – reproduction. Thus the cell represents the limit beyond which further division will result in the extinction of life, and is the smallest thing that can be called 'living'. Unlike whole organisms, which are composed of collections of organs and tissues, or tissues which are composed of billions of cells, cells are not themselves composed of yet smaller sub-units capable of reproducing on their own (although cells contain other, smaller parts, of course). A cell is a single, functional unit like a car, not an aggregate like an army. Taking the wheels off a car does not result in a smaller car. It results in a wreck.

Because cells are so basic, a great deal of biological research is concerned with the study of individual cells and how they work rather than of whole organs made up of billions of cells. We too shall look at single cells, rather than whole people, in much of this book.

Of course, there is no definite rule about how many cells make up an organ or an individual. The human body contains trillions, but a mouse contains far fewer, and a flea fewer still. Can we say what the minimum number is? Very easily, as the smallest number of cells we see a living organism to have is just one. The organisms with only one cell are complete but microscopic in size (usually, anyway, although such one-celled creatures as the amoeba are just visible to the unaided eye). A major type of single-celled organism are the bacteria. These organisms can turn waste into compost, rot food and occasionally cause disease. They are found everywhere – in the air, in water, even living inoffensively inside other, larger organisms. Nearly all are entirely harmless, and many are directly beneficial. Because they only have one cell they are correspondingly simpler than animals like us, with our vast concourse of cells, and so are often chosen as the subject of scientific research. We shall be meeting bacteria many times again.

Bacteria are not typical of the living things with which we are familiar. Most plants and animals contain more than one cell in each organism, from the few dozen in the tiny plants that often form a bright green skin over still ponds to trillions in man. When many cells of the same type come together the result is usually a tissue, a part of the body that is uniform in appearance and function. Bone is a tissue, as it contains only a few types of cells working with one purpose. So is muscle. An interesting corollary is that plants have tissues, too; this is something that will be of practical importance when we discuss how to engineer their genes.

The cells of larger animals and plants have specialized roles to play in their bodies. Each cell is an individual unit, like a house in a village, but they all contribute to the whole by giving some unique resource or specialization. As a village carpenter obtains loaves from the baker and linen from the draper, giving furniture in return, so the skin protects the muscle and bone from the environment while the muscle moves the body and the bone holds it erect.

The inhabitants of this body-village have a few other resemblances to other, more mundane townships. While a lone crofter may sow, spin, bake and repair thatch, few city-dwellers have such universal skills, sacrificing their ability to milk goats for greater expertise in accountancy. So with cells. Some bacteria can live as loners in the cell world, but the cells of such vast cities as the human body need the support of other cells to survive. Only when scientists supply special conditions for them in the laboratory can human cells thrive outside the human body.

To be able to perform these cooperative endeavours animal cells need to be able to communicate, both to pass on the products of their various specialist labours and to inform each other of what their status is and what they need next from other cells. This poses a problem because the cell is to an extent an autonomous unit. Like a house in a village, it is surrounded by a barrier. In animals this barrier is a thin, flexible sheet called the 'plasma membrane'. Plant cells have an additional thicker and more rigid wall outside their plasma membrane called the 'cell wall', as do bacteria, although these two walls are of very different construction. The wood used for carpentry is almost entirely a collection of empty cell walls. In all cases, however, the living cell is surrounded by a barrier which keeps the contents of the cell in and the rest of the world out.

The cell keeps in touch with its neighbours through the plasma membrane much as a human communicates from inside the walls of a house: by putting doors and windows in the walls, through which materials or information may be passed. However, the 'doors' in the plasma membrane are very selective, acting rather like a combination of door and doorman in letting only certain materials in or out. Thus these 'doors' can link all the cells in the body without abolishing the distinctions between them. This selectivity is essential to the functioning of the delicate molecular machinery inside the cell. If every molecule around could get into a cell to jam up the works, or if bits of the molecular clockwork could get out, leaving a hole in the mechanism, the result would be a dead cell.

This picture of the cell as the container of a mass of complex machinery belies our metaphor of the cell as a house, suggesting rather that it is more like a factory. And like a factory, it is organized from a central office where the files are kept. The information needed for life is known to us as a collection of genes, and these genes are coded in the order of bases in DNA. Animal and plant cells keep their DNA in a separate compartment within each cell, called the 'nucleus'. Each cell has a nucleus (except red blood cells), and it is from the nucleus that the rest of the cell receives its orders. Thus each nucleus must contain a copy of *all* the genes, that is, a copy of all the two metres of DNA with which each fertilized human egg is endowed. Considering that the nucleus itself is typically only a few hundred thousandths of a metre across, this poses quite a packing problem for the cell!

We should note that bacteria, unlike our own cells, do not have a nucleus, relying instead on a rather different 'filing system'. However, they do store their DNA in a distinct body in the central region of the cell, although it has no definite edge, unlike our cell's nucleus, and has a rather different chemical structure.

The nucleus is not just a filing cabinet for the DNA, waiting passively for someone to come along and read its contents. It is indeed like a central office, because the information encoded in

DNA needs to be active. DNA transmits its message to the cell, forcing it to listen. This active role is crucial to the cell, as it means that the nucleus is the ultimate source of all decisions and changes of direction in a cell's existence. Scientists can remove the nucleus from some types of cell without immediately breaking the cell into pieces. The resulting 'cell' continues to exist as a cell-shaped body, doing many of the things it was doing before its nucleus disappeared. But it is incapable of change or of growth, the two most characteristic features of life. Without the nucleus the cell is not really alive.

So in our search for what genes are we have tracked them down to a message coded in a molecule of DNA, and found the DNA in the nucleus of the cell.

In fact, this does not explain much, as we have simply changed the problem of how Granny got her red hair into the problem of what the DNA in the cell nucleus actually does. But we have made some progress because, as we mentioned earlier, single cells are much easier to study than whole people, and the single molecule of DNA is easier to study than red hair. Scientists now know a lot about single cells from studying those one-celled organisms, the bacteria.

The DNA must have some specific mechanism by which the information coded in the sequence of bases is translated into action. Just as a person who says that anyone who uses their head can crack a coconut does not mean it literally, so the cell does not use its central store of plans in the nucleus actually to perform all the myriad activities that a cell needs to perform. Instead the nucleus governs from afar, providing the cell with the information required to produce other molecules better suited to making bones and eyes and hair than is DNA. The most prominent among these molecules are the proteins. Proteins are a vital part of life; without them to perform its actions in the cells, all the information-packed DNA in the nucleus would be useless – a dusty library of facts with no readers. Yet the proteins are equally dependent on DNA because, as we shall see, DNA governs exactly how each protein is to be made. Thus proteins and DNA together make the central, indissociable core of the molecular machinery of life.

Proteins are similar to DNA in that both are large molecules made of a large number of smaller units joined together in a chain. The links in the protein chain are amino acids, and they are linked together by a specific type of chemical bond called the 'peptide bond'. Thus short strings of amino acids, very small proteins, are often called 'peptides', and longer strings of amino acids, the proteins proper, are sometimes called 'polypeptides'. Twenty different amino acids are used in most proteins and, unlike the bases in DNA which are chemically rather similar, the amino acids in proteins are chemically very diverse: that is to say, in similar chemical 'surroundings' they will react in different ways. There are a huge number of ways in which we can assemble even a short chain of twenty amino acids, and each will have different, specific chemical properties because of the various chemical properties of their component amino acids. Thus there are an enormous number of possible protein molecules, which can have an equally large range of chemical characteristics.

Despite their wide range of chemical properties, proteins do not share DNA's ability to replicate itself. So whenever the body wants to make a new protein molecule, it must make it from scratch. This would be easy if protein was just a long string of amino acids strung together in any order. But it is not: the chemical properties of the protein depend on exactly which amino acids are strung together and in what order. Thus when it makes a protein the cell needs to be able to put exactly the right amino acids in the right place according to a particular plan, as otherwise it would end up with a protein completely different from the one it needed to make.

The necessity for a plan brings us back to DNA. The information held in one type of gene is simply that which is needed to tell a cell how to make a particular protein. It is a plan of the protein, a plan coded in the genetic code. These types of genes are therefore segments of DNA (almost invariably part of a larger DNA molecule) in which the sequence of A, G, C and T tell the cell how to put a protein together.

The mechanism by which the cell makes the protein, using this information, is complicated, but its outline is simple. First the DNA is copied on to another molecule, RNA (standing for ribonucleic acid). RNA is chemically very similar to DNA: it is a molecule made up of a sequence of bases linked into a long chain. However, unlike DNA a typical RNA molecule consists of only one such chain, not two chains wound round each other in a double helix. Because the double helix structure is central to the mechanism by which DNA duplicates itself, this suggests that RNA cannot duplicate itself in the same way; indeed nearly all RNA is not duplicated from other RNA but is made instead by copying the order of the bases in DNA. Typical RNA molecules are quite short compared with DNA molecules, and only contain the information for one or two proteins. This RNA acts as a messenger, carrying the sequence of bases in the genes from the nucleus, where the DNA is found, to the rest of the cell – consequently it is called 'Messenger RNA' (mRNA for short). It is in the cell outside the nucleus that the mRNA is 'translated' into protein by a huge ensemble of other proteins and RNAs which decode the genetic code to direct the synthesis of a new protein. Thus the synthesis of a new protein involves many other molecules, unlike the synthesis of new DNA which, ultimately, needs only an enzyme and the original DNA.

There is another important difference between protein and DNA besides the number and types of links in their chains and how they are made. While DNA is an extended, straight molecule – two chains wound round each other like stiff wire which is rarely sharply curved – the amino acid chains of proteins fold around themselves in an asymmetric heap. If DNA resembles stiff wire, protein is more like damp string, a small compact blob.

But the blob is of a very special shape, because the protein chains do not clump up at random. The reason they mass together at all is that the amino acid links fit together in some ways much better than in others, and so the protein folds around itself to make the largest possible number of 'good' (i.e. low-energy) contacts between the amino acids. In any one particular type of protein the order of amino acids will always be the same, and so the exact way in which we can best fit those amino acids together in a compact blob will also be the same. Thus the shape of the clump that it forms will always be the same for each molecule of that particular protein. Another protein with a different order of amino acids would form a different-shaped clump. Thus, far from being a random assembly of atoms, the shape which a particular protein assumes is determined by the order of the amino acids in the protein itself.

Of course, it is easy to *say* that this is so. Proving that it was so won Christien Anfinsen and his two co-workers, Stafford Moore and William Stein, a Nobel Prize in 1972, and even today scientists are at a loss to explain why any one protein folds around itself in exactly the way it does. As a result, scientists cannot argue in reverse, and say that they want a protein of a particular shape and that the order of the amino acids in the protein must be exactly *this* . . . It is possible to analyse existing proteins, but not to design new ones.

Although the chemical reasons for a protein's folding pattern may be too complicated for scientists to unravel, the result is simple. If a gene holds the order in which the amino acid units are to be arranged in the protein chain, and if this order determines how the protein folds around itself and hence what shape it will assume, then should the cell need to produce a protein of a particular shape it has only to use the information present in a particular section of DNA, a particular gene whose order of bases specifies the order of the amino acids in that protein, and the correctly shaped protein will result every time.

Shape is very important, especially when coupled with the chemical properties of the amino acids, as that shape must be related to what the protein can do. A spanner has a nut-shaped hole at the end of a long shaft because it requires both a hole that will fit the nut and a lever to exert force. Conversely, *any* device for tightening nuts must have a hole of some sort and a method of exerting force. The same general rule applies to everything. Stools have at least three legs because a stool with only two legs would fall over. Cars have wheels because a car with skis or wooden legs would not move on a road. The same general rule is true of proteins: every protein has a shape tailored to the job it has to do.

The roles which proteins play are many and varied. Some types of protein act as structural members of the body. An example is collagen, a protein important in bones as a reinforcement of the otherwise brittle bone structure, and in other tissues as the foundation on which the cells sit. Because it is a structural material intended to hold things together, the collagen molecule is a long, stiff, fibre-like rope, with units at the ends for connecting it to other collagen molecules. Thus they can slot together in long, girder-like structures of great mechanical strength while remaining flexible. It is a pivotal protein in determining the shape of our bodies, because if it were not for our bones and ligaments (the latter are also composed largely of collagen) we would collapse into shapeless heaps. Indeed, a defect in one of the genes for collagen (there are many genes as there are many different types of collagen, all differing slightly in their order of amino acids) causes the severely disfiguring disease osteogenesis imperfecta. Only one of the body's battery of collagen genes is affected, but the result often deforms the victims so badly that they die within a few years of birth.

Not all shape is determined by the bones. Muscle is another major shape-determining tissue, and in muscle too most of the 'solid' matter is protein. The proteins in muscle are different from those in bone, as their intended function is also different. The muscle proteins can tense up, pulling in the ends of the muscle cells and so making the whole tissue contract: their shape reflects this role, as together they build a structure rather like the wheel-and-ratchet arrangement of a funicular railway.

In addition, more subtle forms of body shaping may be determined by proteins. Cells are not usually just round blobs. Cubic cells, brick-shaped cells, flat or tube-shaped cells are all found in the body, and each has its particular shape in order to best fulfil a particular function. Nerve cells must be long and thin like wires to connect up with other nerve cells all around the body in a giant signalling network. Stomach cells are squat and toughened on one side to resist the action of acid in the stomach.

The shape of a cell is determined at least in part by the action of proteins, which form a network of cables and rods inside each cell to hold it in shape and enable it to move around, as our skeleton does on a larger scale in our bodies. Indeed, this network inside the cell is called the 'cytoskeleton'. How these proteins

G F A F —3

perform this feat is almost completely unknown. Again, it is probably connected with their shape and hence with the order of their amino acids, as only a few very particular proteins are used to make up the greater part of this cytoskeleton. However, dozens more are attached to it in order to link it up with other parts of the cell like the nucleus, and there are maybe a hundred or more undiscovered proteins which alter the precise shape of the whole structure. As we do not understand the functions of these proteins in any detail, or how they are carried out, it is not surprising that scientists have only the haziest notion of how proteins determine the cell's shape. However, it is clear that proteins are the key to how a cell comes to be the shape it is, because many treatments which disrupt the proteins' functions also disrupt cell shape, while other treatments which do not affect proteins do not alter the shape of the treated cells.

You will be able to see how a pattern is building up. The genes control the shape of the proteins; the shape of the proteins is believed to determine the shape of the cell; and the shape of the cell determines how the cell fits into the complete organism to make one specialist unit in the whole. The connection between the genes and the human being is coming a little nearer.

Another important aspect of the usefulness of proteins to life is the cell's ability to put them exactly where it wants them. A collagen molecule is no good inside a cell – it needs to be outside, so that it can link up with a vast network of other collagen molecules to form the resilient base in bone. In recent years it has been found that proteins contain specific signals which tell the cell where that protein is to go, a 'luggage tag' tied on to one end. Thus proteins which are to be 'exported' from the cell have a short string of amino acids, called the 'signal peptide' (recall that 'peptide' is just another name for a string of amino acids) which informs the cell of the protein's destination. All proteins which are to be actively sent out from the cell (a process called 'secretion') have such a signal peptide, although the exact order of the amino acids is not the same in all of them. Another class of proteins will be inserted into the plasma membrane itself, for example to form the 'windows' and 'doors' of the cell we referred to earlier (page 26). These have a different luggage tag, a different signal peptide on the end.

This sorting is quite automatic. The signal peptide is just a string of amino acids on the end of a protein. Because of the particular amino acids used in this short string, they are attracted to other proteins in the cell which direct the whole ensemble – protein plus its luggage tag – to the correct destination. This is an entirely chemical interaction, and does not require some master controller or overseer to make sure everything arrives in the right place. Thus the sequence of amino acids in a protein not only determines the protein's shape, but also where the protein will end up. (Interestingly, once the signal peptide has done its work and got a protein on to the conveyer belt that takes it to its final destination, it is usually cut off the end of the protein so that it does not interfere with the folding-up of the rest of the molecule.) And it is the gene which determines the order of the amino acids. Once again, we come back to the gene as the director of the cell's machinery.

There is one major piece missing from the picture so far. The shape of bones is important, but it is not of much relevance if the limbs never move. Similarly, the existence of a lot of proteins inside the cell, holding it in a particular shape, will be of little use if their function is only to keep the cell alive. A spanner is an ideal shape for an instrument to tighten nuts, but it will never tighten one unless picked up and used. In short, as well as existing, the molecules of the cell must also *act*.

Proteins are the primary agents of *action* in the body.

Most types of proteins do not act as structural material for the body. Instead, they perform more subtle functions. These functions fall into two broad classes: catalysis and regulation. We shall meet the regulatory proteins at the end of this chapter, but let us start with the proteins of catalysis. A catalyst is anything which speeds up a reaction without being consumed by it. In this case the catalysts are a class of proteins called 'enzymes', and they speed up a very wide range of chemical reactions. As these were the first proteins to be identified and certainly have the most spectacular action when separated from the cells in which they are made, we shall look at them first.

Enzymes are proteins whose shape may better be compared with that of a pair of pliers than of a spanner, as they can all latch on to things and pull them apart or twist them together. The 'jaws' of the enzyme proteins are parts of the molecule so shaped that they fit one sort of target very exactly, but any other sort not at all well. Being molecules, their targets are not nails or bits of wire but other molecules, and so enzymes only latch on to particular molecules. Like pliers pulling a nail out of wood or twisting two bits of wire together, the enzymes can join up the molecules they latch on to or break them apart, forming new molecules and hence new materials. This process is very fast, for although each enzyme can break or make only one type of molecule, it can do that exceedingly efficiently. The enzyme molecule is not altered in the process – it emerges from a molecule-twisting session just as it went in – and so it is acting solely as a catalyst. Enzymes were first identified by the discovery of something in damaged cells which could catalyse chemical reactions with great efficiency and great specificity, speeding up a very few selective reactions. This specificity is essential for the cell, of course, as it does not want all its molecules split up and joined together at random in a madman's junk-heap of molecules. But it does mean that for every reaction the cell wants to bring about, for every molecule the cell needs to turn into another molecule, there must be a specific enzyme available to do the job.

To make up for their lack of general ability, enzymes are very effective catalysts. For example, the enzyme rennin (sometimes called 'rennet') breaks open the peptide bond in proteins, so opening up the protein chain to form smaller molecules and ultimately making a pool of separated amino acids. The rennin enzyme is found in calves' stomachs, and its action there is to break up the proteins in the calves' food, including milk. It can also be used to break up the proteins in milk outside the stomach. The milk proteins usually form fairly compact globular shapes which do not tangle up with one another. However, when rennin begins to cut these proteins up, the shorter polypeptide chains which result do not form such neat globular shapes but become tangled up in a large web of bits of protein. The result is that the milk becomes full of tangled bits of protein, and curdles. In the absence of bacteria (which would otherwise turn the milk sour) the proteins would eventually do this on their own. But it would take millennia for anyone to notice the difference because proteins only break up very slowly if left alone, and in the meanwhile many other chemical changes would have been taking place – turning the fats in the milk rancid, the sugars into vinegar, and so on. A few thousandths of a gram of rennin can bring about this change in the proteins in a few minutes, while not speeding up any of the other chemical changes.

What processes in life are affected by these biological catalysts? Are they just curiosities helpful in making cheese, or are they universal to life?

The answer is that they are universal. Every process in the cell is under the aegis of an enzyme. The self-replication of DNA happens only because an enzyme, called 'DNA polymerase', speeds up the operation to a useful rate. Without the enzyme the duplication of DNA would occur on its own, and indeed it is believed that this is how life originally began, but the duplication of all the DNA in a single human cell would take enormous lengths of time – far greater than the life-span of a human being. The production of RNA needs another enzyme called 'RNA polymerase', and the synthesis of protein requires a whole battery of enzymes which make up the complex of proteins and RNA alluded to above. This is why genes are 'read' only inside cells, and why it is so important to keep all the contents of a cell confined within it. Because the most crucial of those contents are the protein molecules which are the enzymes: without a vital enzyme the whole machinery of the cell would grind to a halt. And this in turn is why the integrity of the plasma membrane around the cell is of such importance, and why the cell communicates with other cells only through special 'doors' which allow just a few types of molecule out. These 'doors' are designed not merely to let out or let in some particular molecule, but also to keep in all the enzymes that make the cell work.

The 'doors' are in fact themselves a type of enzyme. They latch on to whatever molecule they are meant to allow in or out of the cell and then alter their own shape to provide the molecule with a pathway into the cell's interior. Sometimes they can even pump molecules one way or the other by hooking themselves into the basic power-supply of the cell. It is just such a pumping mechanism that despatches proteins with a signal peptide to their correct destinations.

Although our cells can make an extraordinary range of enzymes to carry out nearly all the chemical reactions they need to perform, they cannot make every enzyme they could possibly need in all circumstances. There are many chemicals made by our chemical industries, for example, which are quite poisonous, but which our cells cannot destroy. Thus our bodies cannot make the chemical dioxin into a harmless compound (although it would be very useful if they could) because we do not have a suitable enzyme to do so. Similarly, we cannot make an enzyme that will catalyse the chemical reactions which make the compound vitamin B-12. We must therefore avoid dioxin at all costs, but conversely we need vitamin B-12 in our diet. If we get too much of one or not enough of the other, we become ill. Some bacteria do not suffer so many of these metabolic weaknesses and can grow on almost any food, having within their single cell enough genes to make all the enzymes they might need to turn that food into bases, amino acids and all the other molecules necessary for their growth. Many bacteria need vitamin B-12 just as we do, but they have a collection of enzymes for making it out of other, more common molecules; if they are not getting enough vitamin B-12 in their food they can simply make it for themselves. This multi-talented ability to make complicated molecules out of simple ones renders bacteria potentially very useful to us, and it is all because bacteria have those particular genes whose sequence of bases encodes the sequences of amino acids in the enzymes we need but lack. In principle, if we could 'tune up' a bacterium so that it made more vitamin B-12 than it needed, then the bacterium would end up with a surplus of B-12 that we could use for our own consumption.

This is not entirely hypothetical. Plants do not need vitamin B-12, and so they do not make their own, like bacteria, or get it from other living things, as we do. Most people get their B-12 from meat or dairy produce, but vegans — vegetarians who eat no animal produce at all — can get no vitamin B-12 from these sources. Yet they show no signs of deficiency of this vitamin. Why? Because some bacteria which live on the plants and in the soil need B-12, and make their own. The bacteria stick to the plants, and so the people who eat the plants usually eat a few of the bacteria too, and with them all the vitamin B-12 they need. Thus vegans, although they get nearly all their food from plants, obtain all their vitamin B-12 from bacteria. In effect they are using the enzymes from the bacteria to make their vitamin B-12 for them.

What has all this to do with genes?

Enzymes are proteins whose unique shape allows them to latch on to particular molecules and act as catalysts in turning these into other molecules. Proteins are chains of amino acids strung together in a unique order which determines how the molecule folds around itself, and the order is itself determined by the order of bases in a gene. So the bacterium can make vitamin B-12 because of the order of bases in part of its DNA. We cannot make vitamin B-12 in our bodies because our DNA does not contain this information, these particular genes. If we want to make vitamin B-12 we could do so chemically, or use enzymes extracted from a bacterium, or make use of the bacterium itself (as vegans do), but the most efficient way could be to go straight to the gene and use that.

Enzymes are potentially enormously useful. Apart from making vitamins, we could in principle find an enzyme to catalyse any reaction, such as breaking down industrial effluent into harmless waste or dissolving oil-slicks on the sea. If we wanted a new enzyme, or more of an old one, we could make it ourselves by joining the correct amino acids together in the right order, using chemical reactions in a test-tube (although, as we mentioned earlier, finding out what the right order is can be an insuperable problem). Or we could extract it from a living organism, as rennin is obtained from calves' stomachs. But enzymes do not last for ever, and when that batch was exhausted we would have to go back to our test-tube or our calf for some more.

If, on the other hand, we took the DNA whose base sequence tells the cell how to make the enzyme we need, then we could use the self-duplication of DNA to make more of the DNA whenever we wanted, *and* use the DNA to make the enzyme. Indeed, we would only have to start off with one molecule of our DNA and duplicate it until we had as much as we wanted. That DNA could then be used to make an effectively unlimited supply of both enzyme, and more DNA.

In practice, of course, the DNA we started with and the enzyme we produced would both have to be inside a cell, as the enzymes for making proteins and for catalysing the self-duplication of DNA must be held together inside a cell for a useful amount of protein to be made. For convenience this cell could be a bacterium, whose DNA-duplicating enzymes would produce our new piece of DNA as the bacterial cell grew (bacteria can grow very fast indeed), and whose other enzymes could make the enzyme we wanted from the DNA. This, of course, is a typical example of genetic engineering.

Using DNA to produce an enzyme when it is really the enzyme itself we want might seem needlessly circuitous at first glance. But it makes good sense, because we can use the self-duplication of DNA to keep the supply topped up.

And we need not be limited to enzymes. Enzymes are extraordinarily useful proteins, but they are not the only type of protein. We could ask a bacterium to make collagen, for example (although it is unlikely to be commercially very useful). Or we could make it produce one of another class of non-structural proteins, the regulatory proteins. This also has great potential use, but to understand why we must first look at what those proteins are.

'Regulatory proteins' is a catch-all phrase, and not one a scientist would readily use. However, it describes a large and varied group of proteins whose function is not to catalyse reactions or physically to hold the body together, but to pass messages in order that the whole complex of the body may be regulated properly. They are part of our control systems. They can pass messages between points as near to one another as different parts of the cell or as far apart as the brain and the feet, but they are all in the communication business. It is perhaps fitting in this age of electronic communication that more such proteins have been discovered in the past two decades than in the entire previous history of biology.

The first class of such proteins to be discovered were hormones. Hormones are molecules which carry signals from one part of the body to another, distant part through the blood. Insulin is an example of a protein that is a hormone, and indeed was one of the first products of genetic engineering. We will go into what it does in the next chapter, as breakdowns in its functions are the basis of serious disease. Molecules such as insulin need be present only in tiny amounts because their role is to carry a message, not to perform an action: in the same way, the signals entering a radio set are tiny compared with the electrical effects they produce. Insulin is a relatively common hormone, circulating in the blood of healthy individuals in easily detectable amounts, but some of the protein hormones are present in extremely low concentrations. Proteins such as growth hormone are only present in blood in tiny amounts because they are extremely potent: in the case of growth hormone a few thousandths of a gram can dramatically alter the rate at which a child grows. And growth hormone is just one of the more common of a whole range of hormones which regulate how fast we develop. Little wonder that these proteins remained unknown for so long.

Other regulatory proteins are present in more restricted areas of the body. The few of these proteins that we know about are probably just the tip of the iceberg; for every known regulatory protein there may be a dozen still undiscovered. Typical of this kind of molecule is a group of proteins called 'neurotransmitters', examples of which are the endorphin peptides. These short chains of amino acids send signals between cells in the brain (hence their name), and the endorphins in particular are believed to be involved in transmitting the signals which tell us whether we are in pain or not; they are the natural equivalents of morphine. Because their function is to carry a signal from a few nerve cells to a small number of other nerve cells, there need only be tiny amounts of these proteins in a whole brain. This is why the neurotransmitters we know about may be simply the most common of a whole battery of such proteins. Evidence that this is so comes from simple organisms such as the sea slug (*Aplysia*) which have only a few hundred nerve cells in their whole body. We know about just as many neurotransmitters in these organisms as have been found in the human brain, despite the fact that they are so much simpler; this is because it is much easier to find a protein produced by one nerve cell if that cell is one of a hundred than if it is one of several billion. Starting with a billion brain cells, it is an almost impossible task to find just a few dozen protein molecules in them, so only those made by many nerve cells at once can be detected. Even if you could find them, you would then not have enough to analyse in order to discover exactly what their function might be.

Why are they such rare molecules? Because they are produced in small amounts by a few cells, yes; but why is that? Ultimately, it must be because the *genes* for these proteins direct that they should be made only in those cells, and in small amounts. It would be more convenient for scientists if the genes could be readjusted so as to produce as large a quantity of these neurotransmitters as the body produces of, say, insulin; there would then be enough for us to find out how the brain uses these proteins, and so come nearer to an understanding of how the brain works both in normal individuals and in those who are ill. But we cannot arrange for this to occur in the brain itself, as otherwise the resulting flood of neurotransmitters would undoubtedly kill the organism. Anyway, the scientist would be attempting to alter one gene in more than 100,000 without affecting the rest, which is rather like trying to make a racing car by shooting at random at a pile of 100,000 Model T Fords. It is clear that another approach was needed to the problem of altering the genes to make more regulator protein. That new approach was created by genetic engineering.

There are many other regulatory proteins which perform a number of different roles. Nearly all animal cells have proteins on their surfaces called the 'histocompatability proteins' – these are responsible for the 'tissue rejection' which makes transplant operations so hard to carry out. They are a means of communicating with the immune system which makes us immune to disease (we shall be meeting this again later), and may also be involved in the control of our normal development. Thus these proteins are an example of signalling from cell to cell. Regulator proteins exist inside cells, too. Here they can affect the activity of other proteins – enzymes, for instance. The synthesis of RNA inside the nucleus is subject to the control of a battery of such regulators. Some of these are being identified now, while indirect evidence leads us to suspect the existence of others. Until they are all identified scientists will have an incomplete picture of how the cell works, and so will be able to conclude only roughly what goes wrong with it in disease, and what to do about it. Again, in many cases we lack enough of these proteins to study.

In our brief discussion of the regulator proteins we have come back to our need for more of them. I have not mentioned how they work because, unlike that of enzymes, the regulator proteins' mode of action is often unknown; more protein in the test-tube would put that right. And, as we mentioned when talking about the production of more enzymes, in the long run the most efficient way to make more protein is by using the gene for that protein. This is the stuff of genetic engineering, and indeed the production of a regulatory protein – a hormone – was the first successful commercial application of the new technology. However, we have not yet exhausted the potential of regulatory proteins because we have only talked about what happens if they work properly.

What happens if they go wrong? The result is invariably disease, and here too genetic engineering has made great contributions. So in the next two chapters we shall discuss briefly the ways in which the body can go wrong and what we might want genetic engineering to do about it, before we discuss how these feats are going to be achieved.

Chapter 4
A Cause for Cancer

We have mentioned that the cell is not just a bag of enzymes, but rather that it can have a definite shape and some internal structure. In a healthy organism the shape and structure of the cell and the activities of all the enzymes in it are carefully integrated so that the whole is much more than the sum of the parts: the wheels of the intricate clockwork of the cell mesh perfectly. But this order is not inviolate. One of the long-term aims of genetic engineering is to understand what happens when the orderly machinery of the cell breaks down, when the cell malfunctions to produce chaos in itself and disease in the organism of which it is a part. One day we may be able to prevent all such changes; today we understand only a few of them.

We will group malfunctions of the body according to the time-scale over which they occur, as this reflects what is going wrong with the cells. We will keep to our cell-based view of life, and mention what happens to the whole organism only as an effect of these changes in the cell. This will enable us to see clearly why the causes and the treatment of various types of disease are quite different, and what genetic engineering could bring to their diagnosis and cure.

The most obvious sort of disease is caused by some external agent attacking the body. Being hit with a club may not seem like a 'disease' on the same lines as tuberculosis or leprosy, but they share a very real similarity in that both are caused by something attacking the body from the outside. In the former case, the attacker is a human being using a 'blunt instrument' which features so prominently in detective stories. In the latter, the attacker is a bacterium. While it is rare for someone to want to attack us with a club, and while the edges of tables or stray rocks we bang into by accident do not 'want' to attack us at all, the bacteria which cause disease make a living by preying on other organisms.

Like all living things, bacteria must find nutrients for their growth. The bacteria that cause disease (and I should emphasize that these are in a very small minority among the bacteria, most of which are completely harmless) are rather like humans in that they need a variety of complex foods to survive and cannot make do with simple chemicals. They obtain this food by feeding on the rich wealth of biological molecules already concentrated in other living things. But for these tiny predators stalking and killing prey does not involve days of travel across the savannah, but rather swimming a distance measured in millimetres through our blood or over our skin to attack creatures of their own microscopic size. For the bacteria that cause disease, these 'creatures' are the cells of our bodies. If we were of the same microscopic size as a bacterium, animals would appear not as single individuals but as vast herds of cells, like bison waiting for the wolves. (Actually, bacteria are so small compared with our cells that a better illustration would be a herd of bison being attacked by hamsters! This is one reason why so few bacteria take up the speciality of causing disease.)

The 'teeth' of bacteria are usually enzymes which can break open the molecules of their targets and release the contents to be further broken down and absorbed into the bacterium. A collection of such breaking and absorbing enzymes became quite famous for the role they played in the development of molecular biology. These are the Lac enzymes of the bacterium *Escherichia coli* (universally called *E. coli* for short, to avoid spraying your listeners every time you mention it). This bacterium is usually quite harmless, although some varieties of it can cause disease. The Lac enzymes are involved in giving *E. coli* the ability to 'eat' a type of sugar, called lactose, which is found in milk. One of these enzymes is involved in making a 'door' into the cell which allows the lactose to enter the *E. coli* cell, another in breaking the lactose (which the *E. coli* cannot use easily) into smaller molecules. Similar things happen to the food in our intestines as we break it down into molecules small enough for our cells to use. Our intestines have a wide range of enzymes to break down different foods, and disease-causing bacteria also have a range of enzymes and other proteins they can employ against us, their targets. They do not have all the enzymes they could ever need for attacking any cell, so these bacteria are rather restricted in the types of cell they can attack. As a result, a bacterium which causes disease is likely to affect only a limited number of tissues rather than every cell in a body.

Despite their enzymic armoury, disease bacteria, and indeed most bacteria which do not cause known diseases, are often vulnerable to attack. Some chemicals in nature are poisonous to bacteria but relatively harmless to us. These are the antibiotics (literally 'against life'), and today antibiotics such as penicillin are a standard treatment for infection by bacteria. Penicillin is actually the name of a group of similar chemicals related to one made by a fungus: scientists have improved on the fungal version for our own use. Others likewise are natural products which man has adapted. When given in the right dose, these chemicals kill the bacteria without harming the animal they have invaded.

Some antibiotics act on enzymes that bacteria possess but animals do not. Like all enzymes, the bacterial enzymes are highly specific catalysts of a particular reaction – proteins that latch on to a particular shape of molecule and 'twist' it to join it to another molecule or break it into two. Some antibiotic molecules mimic the molecules which these enzymes are meant to latch on to, but with the key difference that they cannot be twisted in the same way as the enzymes' usual target. They are like a bolt that fits a spanner correctly, but which has no thread. Thus they latch on to the enzyme and then just stay there because the enzyme cannot process them further. This takes that particular enzyme out of circulation for a time, because while the antibiotic is attached to it no other molecules can latch on to be processed. In this way, these types of antibiotics jam a part of the metabolic machinery. The 'sulpha' drugs work in this manner, blocking a key enzyme in the metabolism of many bacteria. We do not have that enzyme: it is an enzyme which makes folic acid, a molecule that is vital to life but which we, lacking the ability to make it afresh, have to eat as part of our food. Bacteria, the generalists among the cells, *can* make it on their own, using several enzymes (among which is the one capable of being jammed by the sulpha drugs), and so have never acquired the ability to take this molecule from their food. So jamming folic acid's synthesis using sulpha drugs deprives bacteria of folic acid without affecting our own supply.

The malfunctions caused by this sort of external attack are of no great theoretical interest, although the load of suffering they cause is enormous. It is obvious what is going on – our cells are under some external attack. Of greater interest to us in trying to understand how the body works are the other classes of malfunction, where the cell goes wrong due to a failure within itself rather than as a result of being dissolved by a bacterium or crushed by a rock. Like external attack, these internal failures can cause severe disease. They occur on three time-scales.

Those that occur over short time-scales are problems the cell can overcome on its own, or which the body as a whole can successfully deal with even if individual cells die. These correspondingly cause diseases from which we recover after a few days or weeks. The most common disease of this sort is infection by a virus.

It is commonly assumed that viruses are like bacteria – tiny living organisms which attack larger organisms for food. But they are not: viruses are not really alive. At no time in their life cycle do they do what is most characteristic of living things – take simple materials and build them into something more complex. Fertilizers are turned into corn, swill into pigs, cornflakes into people. Viruses do not do this. Instead they are chemicals, hugely complicated and constructed of the same ingredients as we are but no more living than DNA in a test-tube. (Indeed, you can sprinkle the DNA extracted from some viruses on their target cells and the DNA alone will cause an infection, although it is only a pure chemical.) Viruses are generally made of two chemicals we have met before: DNA and protein, although in some viruses RNA substitutes for DNA. But, unlike the cells of our bodies, viruses do not contain all the enzymes needed to make more DNA or more protein; instead, they are virtually a protein box to hold their DNA. Their DNA is correspondingly simple – at its most basic, it just says: 'Build more virus.'

All DNA is equal: the target cell can have difficulty in telling the message in the virus's genes from its own. So when the virus attacks the cell, injecting its pirate genes into its victim, the cell obeys this new genetic command as if it were its own – it starts slavishly to bend its every resource towards making more virus while neglecting its duties to itself and to the rest of the body. Thus viruses act as Fifth Columnists, subverting the body's own workings to make more virus.

This makes them very hard to combat. Fifth Columnists have always been the bane of military men, for obvious reasons. They use the society of the very country they wish to destroy, so that to remove them each individual saboteur must be tracked down and eliminated separately. We cannot usually hope to disable viruses with the same wholesale methods as we used against bacteria, such as treating the victim of a virus with an antibiotic. The viruses are using our own enzymes against us, and consequently anything that affects them will probably affect us just as badly.

However, the body has its own counter-insurgency techniques. It can adapt itself so that it can distinguish an invading virus molecule from a native one, and thereafter will be on its guard against the intruder. This is the process called 'immunization', and it comes into operation when we become immune to childhood diseases, catching chicken-pox once and then (if it has all worked properly) never again. We become immune to bacteria, too, as we do to anything else that our bodies detect attacking us. The same process also comes into play when we are immunized against a disease we have never had. We expose ourselves to a virus that has been disabled in some way, and the body 'learns' to recognize that virus as an enemy. Thereafter if we encounter the real, intact virus our bodies recognize it as an invader and attack it at once. We can also become immune to cells from another human being, if our bodies are exposed to those cells. Some of the proteins on the outside of our cells vary slightly between individuals, and these minor variations are enough to tell our bodies that those cells are not our own. The proteins that play the biggest role in this sort of immune response are the histocompatibility proteins we mentioned earlier (page 40): it is by detecting small variations in these proteins that the body knows that a particular piece of tissue does not belong in it at all, and consequently attacks it – this is 'tissue rejection'.

The way the body does this is very ingenious. Again, proteins are the key, this time a class of proteins called 'immunoglobulins' or, more commonly, 'antibodies' (not to be confused with antibiotics). Certain cells in the blood called 'lymphocytes' make antibodies at a low level throughout life, each one of the billions of cells making a different antibody. If a foreign molecule enters the body, a few of these cells will produce an antibody which, by chance, fits that invading molecule rather well. The antibodies then latch on to the invading molecule, and this sends a signal to the cells which made those antibodies, causing them to divide. Those cells then divide, and divide again, and keep on dividing as long as the invading molecule is still around. The result is millions of the descendants of those particular cells in the bloodstream, and they all produce antibodies which bind to the invading molecule. Once that molecule has been connected up to an antibody, other cells can recognize it and clear it away; after a while the invader is destroyed. However, the lymphocytes are still there, so if that same molecule is seen again then they are ready to produce a lot of that particular antibody at a moment's notice to clear it up. This whole network of cells and proteins causes us to be immune to diseases we have met before, and consequently it is called the 'immune system'.

The invading molecules concerned can be isolated ones like proteins, or molecules in the coats of viruses, or in the walls of the cells of parasites. Indeed many parasites change their outer coats every now and again just to get round the body's memory of what their old outer wall looked like. Malaria is such a parasite. Influenza virus also does this; and when it does, a 'new' influenza virus sweeps through the population, like the Asian flu of the mid 1960s or the great influenza epidemic of 1918. Only when most people have been exposed to the 'new' virus does the epidemic die down: by then their lymphocytes have been exposed to the new-outer-coat molecules, and the cells producing antibodies which can latch on to those molecules have proliferated so that our bloodstreams are full of the right antibody.

Still other viruses hide inside the cells of the body, playing sleeper while the cell's clockwork operates flawlessly around them. They show no foreign molecules at all to the lymphocytes, and so no antibodies against them are built up. However, the disease caused by those viruses is not cured – it can break out again one day when the virus becomes active once more. Herpes and hepatitis-B viruses are both able to perform this trick.

This brings us to the second of our time-scales: the long-term malfunctions. A few of these are known to be caused by viruses. Most viruses cause fast-acting disease – the cells of the body can either throw the pirate genes out, with the help of the lymphocytes, before the cells die, or they cannot. Either way, the battle is over in a few weeks or months. However, a few viruses can lead a strange double life, appearing in childhood as a disease which passes in a few weeks, and then, instead of being eradicated from the body, lying low in our cells for years or decades when they emerge again to cause a quite different set of symptoms. Thus children encounter varicella virus as chicken-pox, but when they have recovered from that illness a few will not have got rid of the virus completely; instead it will have gone to ground, and may appear up to fifty years later as the cause of shingles. Cytomegalovirus, which causes a rare type of cancer, is so effective at keeping out of sight in its ex-victims that scientists estimate that about 70 per cent of the population are carrying the virus around with them!

For modern science these viruses are quite ineradicable, as are those which hide inside our cells as part of their normal method of action, such as herpes, hepatitis and the 'new' disease AIDS. They either fade of their own accord, or they do not. Hope for a cure comes not from the medicine of today but from research into what it is that viruses do when they are hidden away inside our cells, and what signal makes them emerge again. But research runs into many problems in the study of this kind of question, not the least of which is that so few of the victims' cells are 'home' for the virus, and in those few cells there is such a small number of viruses. And while the viruses are keeping a low profile, the cell is actively doing all its tasks in keeping the body alive. So trying to find out what the virus is actually doing inside the cell is like trying to listen to a solo violin in a boiler-house. We need to find a way of amplifying the virus's activity so that we can detect it more readily. This is a very similar problem to the one we encountered in the study of neurotransmitters (page 39) – there is not enough material for analysis. Genetic engineering can supply some tools to alleviate this problem.

Few viruses cause disease on such long time-scales as years. Long-term changes are more in harmony with normal human development, and so the most common diseases to occur over these extended time-scales are not caused by viruses but by failings in our own growth and development. We will look very briefly at a few of the more common of these, not because genetic engineering is close to curing any of them but because of their importance to our health.

The cell is not an isolated unit in the community of the body. It has communication channels with its neighbours so that it can coordinate its own activity with that of the cells around it. Cells do not behave like a crowd, with each cell doing whatever it 'feels like' at the time, but rather like a well-drilled army, a coherent march of individuals making a whole that is more sophisticated than its parts. Sometimes one cell, or a group of cells, does not respond correctly to the signals it is receiving from the cells around it. It either fails to perform the actions it should, or actually does something else it is not supposed to be doing at all. Not surprisingly, either event can cause disease.

In the first class of failure – failure to perform – are those scourges of modern man, diabetes and obesity. Both are believed by some doctors to be concerned with how the body uses one particular control system to regulate the way we use the food we eat. The system is a complex one, but as in many complex chains of command every line of communication passes through a central point. In this case the point is the cells which produce the hormone insulin.

A hormone is a chemical messenger which circulates in the blood to take a message from one set of cells to a distant set (see page 39). There are many hormones circulating in the blood at any one time, carrying different messages to various groups of cells, and a number of these hormones are proteins. Among the proteins is insulin, which tells the cells of many tissues how much sugar is available for them to burn for energy and how much should be stored as fat against lean times. The sugar itself comes from food via the digestive process (a type of sugar, glucose, is the one with which the insulin system is concerned). The message is sent out by a particular type of cell which monitors the level of this sugar in the blood.

Two types of failure can occur in a system of this kind. Either the cells that are meant to produce the message, or the cells that are meant to receive it, fail to do so. Both sorts of failure occur in the insulin system.

Failure to produce insulin causes the disease diabetes. For some still undiscovered reason, the cells that are meant to produce insulin when the level of sugar in the blood rises can simply stop doing so, with the result that when a diabetic eats a meal all the sugar from the digestion of that meal goes into the blood and stays there – the cells of the body are relying on the insulin signal to tell them when the sugar from the meal has arrived, and the message is not delivered. All the varied symptoms of diabetes stem from this initial failure. The symptoms range across a wide number of tissues, because many types of cell use sugar as a fuel and can be damaged by an incorrect supply.

There are two potential treatments for such a failure. The disease symptoms result primarily from the accumulation of sugar; consequently, patients can be treated by going on a diet containing very little sugar (or, more exactly, whose digested products contain little sugar). Mild diabetes can be treated successfully in this way. However, this only works if the system has been only mildly disarranged. In more severe cases the excess sugar problem is so bad that it occurs no matter how little of it we eat. So we must use drugs to exhort our cells to make more insulin or, if they have given up entirely, supply the vital insulin ourselves. Usually this means injecting insulin several times a day. Although injecting yourself with insulin like this for several decades can hardly be called a cure for diabetes, it is the best available treatment in severe cases.

Some doctors believe that the insulin system is also linked to obesity. The hormone has two messages to carry. It must not only instruct cells to burn up sugar as a fuel but it must also tell a more restricted section of the body, the fat cells, to store as fat the energy represented in sugar molecules. In severely obese individuals (not those suffering simply from middle-age spread) the cells that store fat seem insensitive to this second message. Instead of moderating their activities they store away as much fat as possible, taking every bit of sugar out of the blood that they can. Consequently they store far too much fat. Some scientists think that these two points are connected, and that these cases of severe obesity are a failure at the reception end of the insulin communication system. The normal amount of insulin is available to pass along a moderating message, but the fat cells seem incapable of receiving it and so continue to store fat like misers who believe the rainy days will never end.

It is probable that there are many causes of obesity, and that abnormalities of the insulin communication system are not the primary cause. Many people can eat as much as they like and never put on weight, to the annoyance of others who have to be more careful, but there is no evidence that their insulin system is different from anyone else's. So there must be at least two factors involved. However, we cannot even treat the insulin problem we do know about successfully by giving or withholding insulin. Injections of insulin would be of no use – the problem is not the amount of insulin present, but its lack of effect on the fat cells. Even a surgeon could not replace all the fat cells with more effective ones. So the only treatment available is fasting–for up to four weeks! Victims may long for an alternative, but until we know more about the cause of such problems there is unlikely to be one.

Failure of a hormone's action can be bad, but too much action can be just as bad. Both types of malfunction are known in the case of growth hormone. As we mentioned earlier (page 39), this hormone tells our bodies to grow when we are of the right age to do so; consequently, a lack of it means that sufferers do not grow properly and so are extremely small. Again, the obvious treatment is to give them a suitable dose of growth hormone. Unfortunately, growth hormone is a rare substance indeed. Unlike insulin, which is easy to obtain from animals, the only source of growth hormone which can be given to human patients is human pituitary gland, a commodity that is understandably in short supply. The reason is that the sequences of amino acids in growth hormone in animals differ slightly from one another, so that only the human version is fully effective in human beings. There is the potential for a severe shortage of human growth hormone, which cannot be alleviated by better methods of obtaining it from the raw material. We need another source of human growth hormone than human beings.

The converse of too little growth hormone is too much. Growth hormone acts when we are growing – when we stop growing the amount of this protein in our blood, never very high to start with, decreases. However, in the disease acromegaly there is too much growth hormone around, and although the amount drops when normal growth is over, it does not drop far enough. When the patients are growing this is not a major problem as they just grow bigger and faster than their peers, but by the time they reach adulthood they are very much taller than average. Unfortunately, unlike other children they do not stop growing then. They do not get even taller, but their bones keep on thickening and their joints become heavier as the bones grow in mass, giving them a heavy, lumbering appearance, wide clumsy-looking hands, huge jaws and 'cavemen' brows. Acromegaly has starred in several spy movies: try to think who the symptoms above might describe before looking at the answer on the next page.*

The treatment for acromegaly is to use a hormone whose effect is the opposite of growth hormone's, a short peptide called 'somatostatin'. However, this is another very rare hormone, produced in only a small area of the brain, and it is hard to obtain in the quantities needed to treat an acromegalic throughout life. Again, what is needed is more of the hormone.

Breakdown in the communication between cells is responsible not only for these diseases involving hormones. It is also behind one of the major killer diseases of the West – cancer.

A child must grow to become an adult, and an adult must still retain the capacity to grow in order to heal wounds, grow hair, replace skin, and so on. But there are limits to growth. In the first year of life an average baby doubles in length; if it kept that up for seventy years it would be millions of miles long by the time it dies, which is clearly impossible. Thus the level of growth hormone falls when the body's growing phase is over. The size of an individual is determined partly by the size of the cells but mostly by how many cells there are (the same types of cells are of very similar size in different individuals). Thus how large someone grows is determined partly by how many times their cells divide, first to make two cells, then again to make four, and so on. A cell usually divides if there are not enough cells around it to fulfil the requirements for cells laid down in the plans for development of the body. The message about how many cells there are around each cell is carried partly by general hormones such as growth hormone, which signal a 'Cells needed' message to the whole body, partly by more specific hormones such as 'epidermal growth factor' which signals a 'Skin cells wanted' message, and partly by the immediate effect of one cell touching another next to it. Ultimately, of course, all these messages pass through the cells' plasma membrane so that they can cause the genes to make additional protein and extra DNA to build more cells. There are special communication channels in the plasma membrane to convey these signals, and those communication channels are proteins.

This sort of control acts on all cells, making sure that there are enough for the body's proper functioning. The cell's response to a signal from outside will depend on what sort of cell it is. Skin cells are always multiplying slowly – hormonal signals may tell them to grow faster, but they will not stop them multiplying altogether. Liver cells usually do not grow, but can do so if a surgeon, or disease, removes part of the liver. Brain cells never grow in adults, even if there is a need for them. Thus a cell's

* The character 'Jaws' in the James Bond films, played by Richard Kiel, shows these characteristic features. However, real acromegalics are not much stronger than normal people, suffer more broken bones than usual due to their great height, and usually have arthritis unless they are treated. They do not have steel teeth!

response to these signals depends on the cell itself as much as on the signal.

But what if the communication system goes wrong? What if the controls break down, and the cell either does not grow or never stops growing? The former case can cause degenerative diseases like muscular dystrophy, which we shall mention in the next chapter. But the latter case is a far more common cause of disease, because if a cell never receives a 'Stop growing' signal it will just carry on multiplying, and if the body's immune system does not realize that something is up, and fails to destroy the increasing mass of cells, then it will continue until it invades another organ, blocks a vital nerve or blood vessel, or just uses up the body's entire supply of energy for itself. Then the patient dies of cancer.

No one knows how this signalling system goes wrong, but recent research is beginning to give us some clues. As usual, DNA is at the centre of the story.

In the DNA of our cells there are many genes which carry the information to make growth hormones, growth factors, the proteins in the plasma membrane that receive the messages from these hormones, and other proteins involved in translating those messages into action. Growth hormone is just one bearer of such a message, and there are several proteins involved in the reception of its signal. Each of these proteins is coded by the information in a gene, and this gene, like any other, can undergo mutation. Most of the time such a mutation will simply take the gene out of action, but very occasionally one will occur which allows a gene to work at a higher level than before, or to produce a slightly altered protein which does not perform in quite the same way as the original protein. If the gene concerned is for one of the growth hormones or related substances, or for one of the proteins that translate their messages into action, then the altered protein can be *too* effective in promoting the growth of the cell in which the mutation took place. The cell will then produce its own signals to grow, or will ignore outside signals to stop growing. Therefore it will continue to multiply, and that can be the start of a cancer.

About a dozen genes have been identified as potential cancercausers. They are called 'proto-oncogenes' in their normal state, and 'oncogenes' in their cancer-causing versions. One which has been identified is probably that for a growth factor; mutation in this gene can result in the same cells both producing this factor and receiving the factor's message. So these cells provide their own signal to grow, regardless of what the other cells of the body are signalling. Other genes are involved in the way that the signal of the hormones causes its effects in the cell: mutations in these can jam the signalling system in its 'grow' setting even when no hormone is around. Curiously, mutation in just one of these genes does not seem to be sufficient to turn a normal cell into a cancer-cell – it takes mutation in two or more genes to start the disease. The reason for this is a mystery because, although a few of the proto-oncogenes' functions have been tentatively identified, there is still no complete picture of how a mutation can lead to cancer in the first place.

How do such mutations occur? Of course, they happen all the time, but the rate of occurrence is fairly low. However, it can be speeded up considerably by radiation or by some chemicals, both of which are known to break DNA and so disrupt the order of the bases. It is therefore no coincidence that both radiation and those same chemicals have been shown to increase the incidence of cancer. Some rare viruses can also cause cancer: these are the retroviruses (which we shall be meeting again in Chapter 10). Sometimes retroviruses disrupt the order of bases in genes by putting their own DNA in the middle of them, sometimes they actually carry an oncogene into a cell from outside.

Thus there is no single cause of cancer. Radiation, chemicals, occasionally viruses, and just random mutation can set it off. It has been suggested that all the proto-oncogenes acted on by these agents produce proteins which operate through a central control system. If scientists can identify such a central control point, they may be able to design a drug which can 'turn it off' and thus make any cancer stop growing. That, however, would be very much for the future. Until we have some better idea of what the proto-oncogenes do in normal cells, and what their mutant counterparts the oncogenes do in cancer cells, such speculation will not lead to practical results.

Between 25 and 30 per cent of deaths in the West are caused by cancer. As with many diseases, the incidence of cancer increases markedly with age. We would expect the number of mutations our cells have picked up to increase as they, and we, get older, but cancer strikes older people far more frequently than we would expect from this reason alone. Our chances of getting many other diseases also increase dramatically as we age. As the whole process of ageing is profoundly mysterious we cannot say why this increase occurs, but one of the long-term goals of the study of cancer is to find out why cancer cells themselves seem to be exempt from this universal law of ageing. While normal cells, and the people who are made out of them, inevitably age, some cancer cells can keep on growing for ever, provided they have somewhere to grow. In the cancer victim, of course, they have a limited life-span – when the patient dies, their cancer cells die with them. But if the cancer cells are removed from the patient and grown in a rich soup of ingredients in a sterile environment (so that no bacteria can get in to attack them), then they can evade this limited life-span and keep on growing. One of the best-known of such cancerous types of cells is the 'cell line' called HeLa, after the patient from whom they were removed, Henrietta Lacks. Henrietta Lacks died in 1951, at the age of thirty-one, of a particularly malignant cervical cancer. Eight months before her death Dr George Gey of the Johns Hopkins Hospital took a biopsy of Lacks's cancer and grew some of the cancerous cells. They are still growing, and have been used in biomedical research in hundreds of laboratories. While Henrietta Lacks, were she alive today, would be sixty-seven years old and undoubtedly slowing down, her cells are showing no signs of ageing at all. (That is not to say they have remained unchanged in the meanwhile: they have accumulated far more mutations in their extra-corporeal existence than normal cells would have done over the same period, so that their genes are now quite a mess!) Normal cells cannot do this – if we remove them from the body they grow for a few months and then run out of steam: an ageing process which mirrors our own. So maybe the discovery of a cure for cancer will be the first step on the road to abolishing ageing itself.

So far, in our brief catalogue of failures of the cell, we

have mentioned only two time-scales, that of months (associated with viral infections) and that of years or decades (associated with failures of communication between cells). One of the most exciting areas of molecular biology is in the diagnosis and even treatment of a rare set of diseases which act on even longer time-scales. These are diseases we can inherit – genetic diseases – and are the subject of the next chapter.

Chapter 5
Unto the Fourth Generation

In Chapter 2 we mentioned mutations, changes in genes which can give the genes' owners new characteristics. Short sight must have started as such a mutation, although nowadays people inherit this 'new' gene from their parents – it is now an old-established gene. But what if a gene which is vital to the health, or even the survival, of an organism suffers a mutation? Then the result would be an altered gene whose effects on those unlucky enough to inherit it would be to make those people seriously ill. Many such genes are known. They are the genes for inherited diseases.

Inherited diseases are not very common, for a very good reason. If someone has such a disease, they have it from birth. If it is severe, that usually means they are ill either from birth or from early childhood. In the days before intensive medical care, the chance that these people would die before they reached marriageable age was high, the chance of finding a spouse who would find them acceptable low, and the probability of an affected woman surviving pregnancy and childbirth was also reduced. So only in the last few decades has it become likely that someone who suffered from such an inherited disease would be able to pass it on to a large family of children. Even today, many inherited diseases are fatal before or during adolescence. So, unlike those for short sight or red hair, the genes for inherited disease cannot accumulate in the population, but rather they only crop up occasionally before this natural selection removes them in its usual, cruelly impartial, way.

Many diseases which we do not think of as inherited may nevertheless be influenced by our genes. Whether we suffer from hay fever, whether we are liable to gout or to depression, is influenced by our genes. But here environmental factors are also at work. 'Inherited disease' usually means a disease to which we will inevitably fall prey, regardless of our environment, if we carry the 'wrong' genes. Having blue eyes would be such a 'disease' if it were harmful. There are many diseases of this sort, and their symptoms may often be related to the mutations which have occurred in the genes. So let us have a look at a couple of examples.

Thalassemia is a disease of the blood. It manifests itself during late fetal growth or early infancy in a range of disturbances in the liver and blood, and without medical help it is usually fatal. There are many different forms of thalassemia depending on exactly which of a battery of genes has been affected, but as the genes concerned are all for one kind of protein we shall treat thalassemia as a single type of disease. Thalassemia is a disease whose inheritance is quite simple. It is caused by a single recessive gene. Thus people with two thalassemia genes will suffer the disease, people with only one (or, of course, with none) will not. The gene or genes concerned produce a protein in normal people, the protein haemoglobin which gives blood its red colour and enables it to carry oxygen from the lungs to the rest of the body. If we fail to produce any haemoglobin our tissues become as starved of oxygen as if we had stopped breathing – hence the severity of the disease.

As with all proteins, the information needed to make the haemoglobin protein is coded by the sequence of bases in a gene. In normal people this gene causes haemoglobin protein to be made when it is required, but in thalassemia victims it cannot. A mutation has occurred, maybe dozens of generations ago, which has rendered the gene useless. The information in it has been scrambled, just as the information in a sentence would be scrambled if a few letters were cut out or redistributed at random. The change, the mutation, can be small – just the change of one base to another base; or large – the removal of the entire gene. The effect is the same: the gene no longer works. In normal people each cell contains not one but two copies of each of our genes. So if just one haemoglobin gene does not work, the other can usually make up the difference and we are not much worse off. But if both haemoglobin genes are knocked out by mutations, then there is no copy left to rely on, and so no haemoglobin protein is made. This is the reason why this mutant gene is recessive. If a cell contains one mutated and one normal gene, clearly the normal gene will take precedence in that it is able to direct the manufacture of haemoglobin while the mutated gene is not. Therefore in this case there is a simple explanation of why one of a pair of genes is recessive, the other dominant.

Most inherited diseases are caused by recessive genes: the majority are due to a mutation which knocks out a gene that would normally produce a vital protein. Osteogenesis imperfecta, which we mentioned on page 31, is another example. This simple picture of dominance and recessiveness breaks down only for one important class of genes, and these are the 'sex-linked' genes. Because of the way these genes are inherited, they behave as if they were dominant in men but recessive in women. (This is because they lie on a chromosome, called X, which is present in the usual two copies in women but only in one copy in men.) So if the gene for a sex-linked disease is spread evenly between men and women, men will have the disease caused by this gene much more frequently than women will. In men, only one copy of a mutant sex-linked gene is needed to cause the disease as it acts as a dominant gene, while in women the same single gene causes no disease as it acts as a recessive gene – women need to obtain two copies, one from each parent, which is a much less likely event.

Haemophilia is a sex-linked genetic disease. Haemophilia, like thalassemia, is caused by the lack of a functioning gene. In the case of haemophilia, the gene is for an enzyme which is part of the complex mechanism allowing blood to clot in a wound. This protein is called 'Factor VIII', and without it blood can only clot very inefficiently. Indeed, a bad cut can be lethal to a haemophiliac, who can bleed to death from it if untreated. Nowadays the treatment is to inject sufferers with Factor VIII, made from normal people's blood, to make up for their own lack (in the same way that we inject insulin protein into diabetics who cannot make that protein for themselves). In Victorian times that treatment was not possible, so haemophilia was a life-threatening disease. Queen Victoria had one haemophilia gene, which caused no problem for her but meant that she passed a copy of that gene on to four of her nine children. For the three women this was as little trouble as it had been for Victoria herself, but the one man (Leopold, Duke of Albany) who inherited one haemophilia gene had the disease. Although Victoria herself was not a haemophiliac, the illness nevertheless cropped up in her male descendants in many of the royal families of Europe.

While haemophilia is quite rare, some genetic diseases are surprisingly common (although not as common as, say, diabetes). The most common among European Caucasians is cystic fibrosis, a wasting disease which primarily affects the lungs and pancreas and which kills most of its victims before they reach adulthood. It is caused by a recessive gene. One in 2,000 births in the United Kingdom are of cystic fibrosis victims: about 300 babies a year. No one knows what causes cystic fibrosis – no one knows what the mutant gene does in normal people but fails to do in cystic fibrosis patients – so there is no way to treat these patients successfully. There is no way to tell which parents will have children who suffer from cystic fibrosis. We can only say in retrospect that the parents of the unfortunate child must both be carriers of the disease; that is, they must both have one of the cystic fibrosis genes and hence can pass that gene on to a child, who could therefore get two copies of the mutant gene and, with them, the disease. There is no good method for detecting carriers, although geneticists can, of course, estimate the chances that someone is a carrier. For example, if you have a nephew with cystic fibrosis but no affected brothers or sisters, the chance that you are a cystic fibrosis carrier is about one in three.

Medical geneticists have become extremely expert at estimating these odds, not only for diseases such as cystic fibrosis and haemophilia, where the possibilities are easy to work out, but also for much more complex problems. In the absence of sure knowledge, patients are glad of these statistical risk estimates; indeed, most large hospitals have departments for genetic counselling set up in response to the demands for them. Six hundred parents a year, with 300 babies, wanting to know about the risks associated with cystic fibrosis alone represent a lot of anguish, and any clues that can alleviate it are as medically useful as antibiotics or cancer drugs. But a good test to detect carriers, so that relatives of victims of genetic diseases can be sure, would be very welcome. Usually the only certain way is to examine the genes themselves because, as we have seen, the effects of a single recessive gene may be masked by its dominant partner. Until the advent of genetic engineering the only way to look at the genes was to observe their effects on our children.

For the incidence of cystic fibrosis to be so high the number of healthy carriers of the gene must be far more frequent, as it is only possible to produce a cystic fibrosis sufferer if two carriers by chance get together to have children. And indeed, about one in twenty-five normal Caucasian individuals in Europe and North America are cystic fibrosis carriers! Mercifully, other genetic diseases are far rarer, but they still result in nearly 2,000 children being born every year in the United Kingdom with a disease that is the result of their inheritance of two copies of a recessive gene. The number of children born with a disease which is caused by a dominant gene is fewer – about 550. About 400 are born with sex-linked genetic diseases, but because of the nature of sex-linked disease, nearly all are boys. About 150 of this last category are sufferers from Duchenne muscular dystrophy, which results in gradual wasting of the muscles so that an apparently normal toddler slowly loses the power to walk as he grows older, usually dying in his mid teens. Again, there is no reliable way of finding out whether a woman is a carrier of the gene until she and her partner have an affected son.

These are the figures for the most terrible genetic diseases, those which cripple or kill in early life. But they may only be the tip of the iceberg.

Few of the known inherited diseases affect people after middle age. The most common is Huntington's chorea, a degenerative disease of the brain. About 50,000 people in the United Kingdom *may* have Huntington's chorea, but they will not all find out until some time in their late thirties to early fifties, when the symptoms start to show; then half of them will find that they have the lethal dominant gene for this disease and will develop the symptoms.

But many more common diseases are also believed to have a genetic component, and these usually strike in later life. Earlier, we mentioned hay fever and gout. Heart disease, diabetes and cancer are more serious examples of environment and genetics combining to produce the disease. Again, people who carry particular combinations of genes may be far more likely to develop one particular disease, given the right (or rather, wrong) environmental conditions, than other people. We have already mentioned in Chapter 4 that there are genes, the proto-oncogenes, whose mutation is a part cause of cancer. However, it *is* only a part cause, and other genes must be involved too. Variations in these genes could make their carrier much more susceptible to the effect of a mutation in a proto-oncogene.

Even someone who suffers from emphysema caused by cigarette smoking is affected by what genes he or she carries. Such a disease might be thought to be caused solely by the environment – in this case the environment of lots of smoke in the lungs. But scientists have found that a particular gene, carried by a minority of people, can make one smoker much more likely to suffer from emphysema than another who in fact smokes just as much: genetics and environment combine.

While being able to detect carriers of the cystic fibrosis gene would be a matter of life and death to a few, being able to detect the genes which make us susceptible to cancer or heart disease could be a useful adjunct to health care for many more. If a diabetic's parents could be told when their child was born that he was at risk of developing diabetes later in life and that they should keep his dietary sugar down to help prevent this happening, such advance warning might prevent the child from developing the disease. The fears of life and death which are part of daily life for families with members who suffer from hereditary disease are reflected in less dramatic form in most of our lives. If Grandad died of cancer at forty, are we going the same way? In principle, proper genetic tests could set out the *real* risks, and put a lot of minds at rest.

This is not to say that people would actually use such tests if they were available, or that they would take any notice of the results if they did use them. Alleviating worry is useful only if there is anxiety in the first place – if some potential heart disease victims are quite happy to live until something kills them and do not really mind what that 'something' is, then it would be foolish to impose a genetic test upon them. But there are many people who would be grateful to have such knowledge, and many thousands of children born every year could live much longer and fuller lives if they knew early enough that they had a predisposition to a particular common disease. They could then use that knowledge, or ignore it. After all, some millions of people in Britain know that smoking will make their shortened lives less active and more unpleasant because of disease, yet they exercise their right to ignore that knowledge. Knowledge can always be ignored if it is there. It cannot be acted upon if it does not exist.

This, however, is for the future, because no one knows what genes predispose us to cancer or heart disease. We simply know that such things 'run in families'. So the only way to detect such genes now is to wait until Grandad dies of cancer at forty and then say, 'Well, he probably had that gene.' What would be more useful would be to detect such genes before their lethal work is done. As we shall see, genetic engineering offers the hope of doing this.

The problems we should like to address with the techniques of genetic engineering are now fairly plain. In the first place, we would like to alter the genes for proteins so that we could use the latter to produce materials – enzymes, for example, or hormones. We would also like to have more of some proteins simply for study, as the genes in our bodies make so few of them – neurotransmitters would fall into this category. Surely there is use for a protein which would be a safe substitute for morphine, but are the endorphins such proteins? Lastly, we would like to use genetic engineering as a tool to study how genes work, or fail to work, in human beings, so that we can detect mutation in those genes in potential disease victims and carriers, and design better treatments based on a full understanding of what is really going wrong. With these goals in mind, let us look at the techniques we might use to achieve them.

Chapter 6
The Grammar of the Message

Before we plunge into genetic engineering itself, we need to clear up two points about the genes we are to manipulate. In Chapter 2, I was deliberately vague about how the information in the DNA, the genes, is used and how the pieces of DNA with that information in them fit together. But both these points are important to the way in which genetic engineering is performed.

The body needs to know when and how to use the databank in its DNA. We must respond with the right enzyme in a split second in an emergency, so some sort of controller system is needed, 'a master signalman' to point to a gene and say, 'This one', when the cell needs that particular piece of information.

Such a controller itself needs to know which genes to start up, and that requires more information. The only databank in the cell is that stored in the DNA, so more DNA must provide the information for the controller: in other words, *genes* must be the controllers of other genes. Thus we come to the paradox that the only way to make genes work on time is to have other genes which tell them when to work and when to rest.

This may seem rather daft as we now need to know what controls *those* genes, and of course by the same logic the answer must be yet more genes – and so on up an endless ladder of genes controlling genes controlling genes. But the argument is not useless on account of this apparent paradox, because there is good reason for saying that such 'controller genes' do exist. They have been found.

The first controller genes were found in a bacterium. Bacteria have far fewer genes than we do: they have fewer genes for proteins, as they do not need to make special proteins for hair, skin, muscle, bone and so on which man must make, and they also have fewer controller genes. Our cells need several levels of control for our genes, some deciding whether the gene should be used now or later, some whether it should be used in this cell or that cell, some whether it should be used in infancy or adulthood. By contrast, bacteria need only the now-or-later level of control. Thus we would expect the genes in bacteria to comprise a rather simpler set of instructions than ours, in the same way that the plans for building a garden shed are simpler than those for building a city. In the city the sewers and gas mains must go in before the houses, manhole-covers must be put on after manholes have been dug, and so on – what you build is no more important than the order in which you build it. You can always cut a hole in a garden shed for a window if you forget to put one in during construction.

This expectation is borne out in practice. People, and indeed most animals and plants, have a much larger amount of DNA in their cells than a bacterium has. We find that each human cell contains about 1,500 times as much DNA as each *E. coli* cell. Thus scientists naturally turn to the simpler system of a bacterium's genes when they want to start working out how those genes act.

The first controller genes were found in *E. coli*. These regulate the Lac cluster of genes, which we met in Chapter 4. They allow the *E. coli* to break down the sugar lactose. Clearly they would be of no use if there is no lactose around, so *E. coli* needs to be able to regulate the work of these genes so that they only make their proteins when there is lactose about for them to act upon. So in the Lac cluster of genes there are three genes associated with making proteins: they are called A, Y and Z. But there is also one other gene, the i gene, and several short bits of DNA next to the Z gene that are concerned with controlling the way in which *E. coli* uses the information in the A, Y and Z genes. When there is lactose around, the enzyme RNA polymerase (which is the enzyme that catalyses the manufacture of RNA, the messenger taking the information from the genes to the site of protein manufacture) latches on to the Lac DNA at a spot called

P, just up from the Lac genes, and begins to make RNA, starting just down from that point. Thus the *P* segment of DNA acts as a 'Start here' signal, like the capital letter at the opening of a sentence. However, if lactose is absent the gene *i* comes into play, because the protein *it* makes now latches on to the Lac DNA at another spot, called *O*, which is between the *P* piece of DNA and the Lac genes. So when the RNA polymerase molecule comes along to start making RNA, it finds the protein made from the *i* gene's information blocking the way, and so no Lac RNA is made. Thus by making a little of an extra protein – the *i* gene protein – *E. coli* can regulate the manufacture of a large amount of the three Lac proteins.

How do we know that all those genes and bits of DNA are there, and are doing what we say they are doing? In the same way as we discovered that so many things in man are influenced by genes: we can find mutations which alter how those genes work. A mutation in the *O* region, for example, a change in the order of bases, would mean that the *i* protein would no longer recognize that particular bit of DNA. Like many proteins, the *i* protein is very pernickety about what molecules it latches on to, and if we change the order of the bases in its 'target' DNA, then it will no longer latch on to it so effectively. Scrambling the order of the *O* region will effectively remove it from consideration in our control system, and therefore *E. coli* will always produce Lac proteins. By carefully unravelling the effects of such mutations and by observing what they do in different situations, scientists were able to work out which bit of DNA did what. The definitive proof of the existence of these controller genes was provided by French scientists François Jacob, Jacques Monod and André Luroff in the early 1960s, and biologists were so impressed by their findings that, as well as awarding them the Nobel Prize in 1965, they named the control system they discovered after its principal discoverers – the 'Jacob and Monod model'.

Having three bits of DNA to control three genes may seem a rather odd excess of Chiefs over Indians. But the cell relies on the genes for *all* its information. It cannot search out the beginning of a gene and start reading at that point without some information as to where the start of the gene is, and the need for that information means another piece of DNA. Of course, you might think that *that* piece of DNA would need another to tell the cell where *it* is, and so on, but in fact the cell gets round that sort of thing by using one gene to locate the start of several other genes. Thus the Lac P segment points to the start of the A, Y and Z genes. The cell can locate the P segment because the enzyme RNA polymerase that makes the messenger RNA is so shaped that it latches on to any piece of P DNA. So all you need to locate a number of P segments is one RNA polymerase molecule, and each P segment itself signals the start of several genes. However, RNA polymerase is a protein, and so there needs to be an RNA polymerase gene, which must have its own equivalent of the Lac cluster P segment . . . there is no getting away from the need for genes.

Thus the genes are rather like a sentence, but with every punctuation mark spelt out in full. The order to 'Make Lac' is therefore translated into 'START-IF-LACTOSE-AROUND-THEN-MAKE-Z-MAKE-Y-MAKE-A-STOP'.

All genes need some sort of signal to tell the cell where the gene's message begins. The Lac P segment is a specific example of such a segment. It was called P for the following reason. Mutations in that region of the DNA shut down all Lac manufacture – this is not surprising, as without the P segment the gene has no 'Start here' signal. Thus this stretch of DNA was found to be promoting the proper function of the Lac gene group, and so was called 'promoter', or P for short. These segments have turned out to be a very important feature of genes: although the specific sequence of bases used in a particular gene may vary, it will almost always have a segment of DNA which has the function of acting as a 'Start here' signal (if it is to work), and the segment is consequently always called a 'promoter'. This, then, is another sort of regulator, analogous in function to the regulator proteins we met in Chapter 3 but with a very different role. Promoters act at the level of genes, and only on the genes in a particular cell – they cannot affect other cells. Indeed, they are more fussy than that. Like the capital which signals the start of a sentence, a promoter has to be right next to the gene whose start it marks. In bacteria like *E. coli*, the promoters for all the various genes are next to those genes. The Lac set of genes, then, are all quite near each other on a piece of DNA: they are the information encoded in different sections of a single segment of double helix. But what of all the other genes in *E. coli*? Nearly all the genes in *E. coli* are on one huge DNA molecule. Thus each gene is just a segment of the double helix, and it is impossible to tell where one gene finishes and another begins just by looking for breaks or ends in the double helix – the only sign is that the order of the bases stops encoding the information for one gene and starts encoding that needed for the next. In *E. coli* this DNA molecule is about 4,200,000 bases long, and its ends are joined together to make a loop.

Bacteria differ from higher forms of life in this basic organization. We mentioned before that the cells of bacteria, unlike those of animals and plants, do not have a nucleus. The way their DNA is arranged is also different. In plants and animals the DNA is in several straight pieces, not in one loop. Those pieces are called 'chromosomes' (or, to be more exact, when they are packed up in a thick coat of protein they are called chromosomes: they are then so big they can actually be seen under the microscope). Each human cell contains forty-six such lengths of DNA – forty-six chromosomes – with a total length of nearly two metres of DNA, about 6,600,000,000 bases in total.

The size of the DNA loop in bacteria varies in different species. Further, one bacterium can have several sizes of DNA loop in its cell. As well as the main loop of DNA, which carries most of the genes, bacteria can hold much smaller loops, 'only' a few tens of thousands of bases long and carrying only a few genes. Such small loops of DNA are called 'plasmids'. They often carry genes concerned with unusual or specialist activities, such as surviving in soil which has been contaminated with poisonous chemicals or combating the antibiotics we talked about in Chapter 4. They act as specialists, giving advice (information) on activities with which the bacteria are not usually involved. Thus they carry only those genes involved in performing that particular specialist activity (together with the DNA which makes sure that they duplicate themselves efficiently when the cell divides), which is only a tiny fraction of the number of genes in a bacterium's cell. If we imagine the total DNA in each human cell as a cassette tape 2,800 kilometres long, then the DNA loops in bacteria are a few kilometres long, while plasmids are a mere few hundred metres or so in circumference – the same sort of length as a commercial cassette tape.

Bacteria are not the only organisms to sport plasmids. Some types of yeast also possess them, and animal and plant cells can contain similar small loops of DNA.

The plasmids are useful to the bacterial cell, as they give the cell some new specialist abilities which it did not possess before, such as being able to resist the action of penicillin. But not all small pieces of DNA are so beneficial. There are others which we find occasionally in cells – the pieces of DNA of viruses. Unlike the plasmids, these DNAs give nothing to the cell. On the contrary, their genes have a simple message: 'Build more virus.' Just as viruses can cause disease in people by destroying cells in their enthusiasm for building viruses, so other viruses can attack and kill bacterial cells. Viruses are far smaller than bacteria or human cells, and if they have just the right proteins for penetrating one of these types of cells, then they will enter it and re-program it to produce more virus. Of course, that penetration manoeuvre is not easy, and so a virus which can penetrate a bacterial cell would certainly not be able to harm human cells.

The genes for viruses share two important features with those of plasmids. The DNA on which they are encoded is quite small, in the range of tens of thousands of bases rather than that of the millions in the main loop of DNA in *E. coli*. Secondly, these pieces of DNA, unlike the main loop of DNA in the bacterial cell or the chromosomes in the nucleus of a human cell, are not confined to one cell only.

The DNA in a virus obviously cannot be confined to one cell. If it were, it would never be able to infect any other cells. When a virus infects a new cell it naturally takes its DNA, its 'pirate genes', along with it.

The DNA in plasmids also may move between cells, but in a different way, when one cell passes a copy of the plasmid DNA on to another – an exchange of 'expert' genes. Unlike invasion by the DNA of a virus, a process that any cell would try to avert, the exchange of plasmid DNAs between bacterial cells is something the cells themselves encourage, although some of the genes necessary to make it happen are nearly always part of the plasmid DNA itself. Cells can also sometimes spot a piece of DNA floating free nearby and, if it is small enough, can take it in and make use of whatever genes it contains. Thus plasmids can be picked up by a bacterial cell even when they are not already in one.

The transfer of plasmids from one bacterium to another is one way in which resistance to a new antibiotic can spread to lots of different types of bacteria soon after a new drug is released for medical use. Before the new drug is used, only one bacterium in billions will have the necessary genes to make the bacterium resistant to it: for the majority of bacteria, such genes would be useless. When the new drug is released those genes suddenly become crucial, and any bacterium which has them or can acquire them flourishes while its siblings die. The chance of a bacterium acquiring a new gene by mutation of an old one are very small, but if they can obtain the gene ready-made on a plasmid their chances of survival increase dramatically. So the plasmid is passed from one cell to another like a new recipe, each cell using the self-duplication of DNA to make its own copy, and eventually all the surviving bacteria have a copy of the new plasmid and so are resistant to the new antibiotic.

This ability of plasmids and viruses to transfer from one cell to another is both a blessing and liability to the genetic engineer, as we shall see in the next chapter.

These facts about the nature of genes and how they are controlled were discovered in the 1950s and 1960s by experiments performed almost exclusively on bacteria. Are these organisms unique, or are the genes of the more complex organisms also organized into clusters, with promoters at one end?

This is rather difficult to decide. The presence of plasmids or viruses is relatively easy to check, but 'controller genes' like the *i* gene of the Lac cluster and short segments of DNA like promoters are not so easy to trace, and the elucidation of how they work is a Herculean task. There is a tremendous amount of DNA in each human cell, and to study the effects of one controller gene among so much DNA is like attempting to study one particular bar of Elgar's music from a jumbled heap of 10,000 scores.

The problem is further exacerbated by the physical size of the DNA. As with any other molecule, the best way to study it is to use chemical techniques, but the particular gene we want to study may only be a few thousand bases long, and is part of a whole chromosome of up to 300,000,000 bases long. Doing scientific studies on such tiny amounts of DNA with the chemical techniques available is like trying to find a needle in a haystack with a bulldozer.

Genetic engineering can get round this problem because it can provide a large amount of a 'pure' gene, that is a short piece of DNA whose bases encode one piece of information in isolation from all the rest of the DNA of the cell. Once such a 'pure gene' is available, it can be analysed in much greater detail than can an 'impure gene' consisting of one gene we wish to study embedded in a whole chromosome which we do not – it is easier to make a drawing of a leaf from one twig than from a panoramic view of a whole forest. Making 'pure genes' is not easy, but it can be done. This is the essence of genetic engineering. Here is how it works.

Chapter 7
Nuts and Bolts

The study of genes and how they work, which we have surveyed in the last few chapters, is of practical use mainly because scientists can manipulate genes, and can take genes from one organism to another. They can *use* genetics, in the same way that the nineteenth-century chemists found they could use the previously academic study of chemistry, and so started the billion-pound chemical industry of today. We have already seen what we would like to be able to make genes do: to make new products and to give us a better understanding of how our bodies function in sickness and in health. To achieve these aims scientists have developed techniques for altering the DNA itself, generating new DNA molecules and hence new genes never seen before in nature.

It is important to look at how this is done if we are to understand the limitations and the potential of genetic engineering. So in this chapter we will outline the basic tools of genetic engineering as it is applied to bacteria. In Chapter 10 we shall extend this toolbox of techniques to methods of engineering higher organisms, including man.

The first problem is one of numbers. A human being contains enough DNA for a million genes, but the genetic engineer wishes to study and manipulate only one gene. A prerequisite for threading a needle is that you first separate it from the haystack. However, the genetic haystack presents some unique problems in needle-finding.

To start with, the DNA is not in a lot of convenient gene-sized pieces. First we must cut our gene out from a long DNA double helix containing thousands of other genes. Then there is the problem of identification. Suppose we had a pair of exceedingly small scissors to cut our DNA into gene-sized fragments. All DNA is the same except for the *order* of the bases (with a few exceptions which are not of much practical use to us); how do we distinguish our gene from all the others? The needle we are searching for has turned out to be the same colour as hay. Clearly, a less direct method will be needed than the use of small scissors! That method is the technology of 'gene cloning': it allows us to both separate a gene from all the other genes in the body and to make a large amount of it.

The first step is to make what is known as a 'vector molecule'. This is a carrier, a supporting molecule which will carry the gene we want into a cell and help it to duplicate itself there. (It is rather like the human carriers of genetic diseases, in that the gene we want can be carried around by the vector without affecting it very much.) The vector is a small piece of DNA which will duplicate itself efficiently in the cell. It has to be small because large pieces of DNA are very inconvenient to handle in the test-tube, and it has to be an efficient self-duplicator because we will be wanting a great deal of it. Of course, all DNA can self-duplicate, but the enzyme which helps it to do so, DNA polymerase, is choosy about which sequence of bases it duplicates. The DNA polymerase needs a 'Start here' signal to begin making new DNA, just as genes need a 'Start here' signal. This avoids the confusion of having two DNA polymerase enzyme molecules trying to duplicate the same DNA molecule in opposite directions, and the vector DNA must therefore contain such a signal if it is to be useful to us.

The two types of small segments of DNA we met in the last chapter, plasmids and virus DNA, are of a suitable size. In addition, they have the property that they can be transferred from one cell to another more easily than most DNAs, and even from the test-tube into the cell. The plan, then, is to join a piece of DNA we want to duplicate in large amounts on to a plasmid or a virus DNA molecule, thus joining it to the 'Start here' signal, and then to put it into a cell where it will be duplicated. Plasmids and virus DNAs have technical advantages for different types of genetic engineering, but the principles are similar for both kinds of vector molecule. Plasmids were the first type of vector molecule to be used, partly because of technical problems in using virus DNAs in this way, and partly because a bacterial cell with a plasmid in it could grow and divide to produce a whole line of plasmid-containing bacteria, while virus-containing cells tend to die rather fast. If we want to clone a gene into a bacterium so that the latter can produce a new protein, it is not very useful if the bacterium dies two hours after we put the new gene into it!

So we have our vector, and we want to put a gene into it. The next problem is to isolate a section of DNA which contains just the gene we want.

For many years this was an insurmountable problem. Not that it is impossible to break up a long piece of DNA into shorter pieces – sufficient mechanical battering would do that. (Indeed, the extremely long DNA molecules found in human cells are so fragile that they will be snapped into smaller pieces just by pouring them gently from one tube into another.) But we do not want merely to bash the DNA to bits: we want those bits to fit the vector molecule like pieces of a jigsaw puzzle, and that means that both the vector DNA and the DNA we want to attach to it must be broken in just the right place. We must also produce the right sort of 'ends', so that the pieces can be joined up again into a harmonious whole.

To do this, scientists use a set of enzymes which rely on the properties of DNA itself. DNA is made of bases ordered in an extremely complex way, the order conveying information to those that are able to read it. But not all enzymes *can* read it. The battery of proteins involved in making new proteins can read the message of the genetic code. But to other enzymes this code is a meaningless sequence of bases, just as a message in Sanskrit is a message, encoded in marks on paper, to a Hindu priest, but is nothing more than a lot of squiggles to the average Briton.

One feature of such random strings of 'meaningless' characters is that small groups of characters will occur every now and again just by chance, regardless of the message they might carry to other eyes. Thus the two-letter combination 'ot' occurs nine times in the last paragraph (ignoring spaces), and you could find that out whether you understood what the words meant or not. In DNA any short sequence of bases will have its own characteristic shape, as the bases are parts of the DNA molecule with *their* characteristic shapes. The basic shape of the double helix remains roughly the same, but the details will vary. The enzymes we use to cut DNA recognize these subtle shape differences in the DNA as saying, for example, 'Here is the sequence GAATTC' (while the enzymes which read the genetic code might read the same bases as: 'Put the amino acids glutamate and phenylalanine next'). The enzymes, when they have found the sequence of bases which they recognize, cut the DNA double helix at that point, making two double helices where we had one before. For historical reasons, the enzymes which do this are called 'restriction enzymes'.

But the restriction enzymes do not just cut the DNA. DNA is a double molecule, two long chains wrapped round each other in a double helix. Clearly, both chains must be broken if we are to make two new double helices from our initial one. But some restriction enzymes do not cut the two chains at exactly the same site. Instead, they cut them at sites a few bases apart, so when the two shorter double helices are pulled apart, there are two short strands of *single* helix sticking out of the end of each. In the GAATTC case, one chain is cut after the G, the other before the C. This means that an AATT length sticks out of the end of each of the two new double helices.

Now you will recall that the bases of DNA have a chemical affinity for their complementary bases, and that if we separate the two chains which make up the double helix, this attraction between complementary bases will pull them back together again. The attraction of complementary partners is the basis of the self-duplication of DNA. The short bits of DNA which jut out from the ends of the DNA which our restriction enzyme has just cut also have bases which are not locked together with their complementary partners. They have no 'other chain' of DNA to latch on to, and in the absence of an enzyme they latch on to free bases very inefficiently. However, there *are* some bases available, and what is more they are in exactly the right order to form a complementary chain to the unpaired bases. Those bases are in the short stretch of single chain which juts out from the other cut end. Because they were made by cutting the same molecule, they must inevitably have complementary sequences of bases. So the solution for the DNA is simple – the single chain sections at the cut ends simply pull themselves together again. Because of their tendency to stick together once more when they have been cut, such ends of DNA with pieces jutting out like this are called 'sticky ends'.

Sticky ends do not bond together very strongly, as otherwise there would be no point in cutting the DNA in the first place. However, if we pull them apart gently, they will tend to come back together again given the chance.

An important point to remember here is that any particular restriction enzyme (and hundreds are known) will always give the same sticky ends. This is because one particular restriction enzyme will 'recognize' one short sequence of bases and cut the DNA at that specific site – whether those bases are part of a gene or a promoter or have some other function has no significance at all for the restriction enzyme. Conversely, whenever we cut a piece of DNA with a given restriction enzyme, we will get pieces whose lengths and sequences may vary enormously but whose ends are always the same, and those ends will stick together again regardless of what is in between them. For both the cutting and the sticking, it is the short sequence of bases that matters, not the bases in between them.

This is just what the genetic engineer needs. Not only do we have a way of fragmenting DNA which does not smash it to bits in an uncontrolled way, but we also know that the DNA will tend to stick on to any other DNA with the same sticky ends. We cannot be sure which restriction enzyme will produce the right sections of DNA. More than a hundred are known, so it is likely that somewhere among them will be one which cuts the DNA on either side of the gene we want but not in the middle, like a pair of scissors which cuts a passage of English at full stops, and so separates it into sentence-sized pieces. Of course, *finding* that enzyme can be tiresome . . .

These DNA-cutting enzymes have an additional advantage. As they cut the DNA only when a particular sequence of bases turns up and not just anywhere, we will always know exactly how they will cut up a given piece of DNA. As a pair of scissors which cuts at the letter sequence 'ot' will always cut the 96th Psalm into eight pieces no matter which Bible we cut up, so we can rely on any given restriction enzyme to cut up the DNA from any particular source in exactly the same way every time. For instance, the enzyme which cuts at the base sequence CTGCAG will always cut the human insulin gene into five pieces. Scientists can use this 'constant cutting' pattern to deduce where the genes are on the DNA, and how best to go about linking them up to other genes.

This linking is the last step, and of course it needs another enzyme. Sticky ends will not join themselves up into a continuous double helix again on their own, any more than two pieces of string will tie themselves together; their 'stickyness' is a purely temporary arrangement. To join those sticky ends together permanently we need the enzyme DNA ligase, which joins two sticky ends (and indeed non-sticky ends) together, making a long double helix out of two shorter ones. The place where the restriction enzyme cut is now exactly the same as when we started: the restriction enzyme cut at only one particular short sequence of bases, and the DNA ligase has joined the broken halves of that sequence of bases back together again. But the rest of the molecule is very different. We have taken two sticky ends and joined them together; but, as we said above, we have taken no account of what they are the ends *of* – what DNAs we cut to make those ends. The source of the DNA does not matter either to the restriction enzymes or to the DNA ligase. If one end was a break made by a restriction enzyme in human DNA, and another was one made by the same enzyme in a plasmid molecule, when we joined the ends together we would have a single continuous molecule of which half was plasmid DNA and half human. Indeed, it is possible to cut and paste any pieces of DNA together like this. We could, for example, take a bit from the Lac gene cluster, the human gene for insulin and a plasmid molecule and join them all together in one big loop so that the Lac promoter was next to the human insulin gene, and the two were inserted into the plasmid. Provided the sticky ends were compatible, that would produce a new plasmid which contained not only whatever genes were in the plasmid originally but also the human insulin gene, the latter under the control of the Lac promoter.

This would be a new DNA molecule, one never seen before in nature. We have taken two types of DNA – vector DNA and some other DNA – and joined them up together. In classical genetics such a rearrangement of genes is called 'recombination' and, although the geneticists never dreamed of mixing the genes of E. coli and man when they coined the word, it has been taken into the new genetics to describe the product of these cut-and-paste experiments with DNA. The resulting new, recombined molecule is called 'recombinant DNA'.

So we have a recombinant DNA molecule, a new piece of DNA with a gene we want joined into a vector molecule. What do we do with it?

This brings us back to why we want the gene in the first place. We have seen two major areas in which we want to manipulate genes. We might wish to use it to produce an enzyme, or some other protein like a neurotransmitter or hormone, in which case the more genes we have in bacteria, the greater the amount of protein we will be able to make. Or we might want to study the gene to see how it works as the Lac genes were studied by means of other techniques, and so the greater the amount of DNA we have the more experiments we can do. In either case, the more the merrier.

Here, clones rear their heads. We should remember that cloning has little to do with plots to produce identical copies of Hitler or to give American millionaires identical children. A clone is a number of individual organisms which have identical genetics, in other words they all share exactly the same genes because they all have the same single parent or ancestor. In this case the clone is of bacteria. As bacteria reproduce simply by splitting in two, cloning a bacterium is easy: we simply separate one bacterium from all the others (with which it could mix or mate) and give it enough food to grow. It will split into two, grow, split again producing four, and so on at such a rate that one bacterium could grow into a quarter of a million in twelve hours. Each bacterium would be genetically identical to all the others, and so they would together form a clone.

Now, suppose we had a piece of recombinant DNA in the starting bacterium. As it grew, its DNA would of course be duplicated so that each new cell had its own copy – this would apply to the new, recombinant DNA just as to all the other DNA in the bacterium. After twelve hours or so we would have 250,000 bacteria, each incorporating a copy of our recombinant DNA molecule. In a clone of bacteria we would have a clone of DNA, a large number of identical molecules.

Large numbers of the DNA molecule, the gene, are just what we wanted. However, we now see that we have to get the recombinant DNA into a cell before we can achieve those large numbers. As we mentioned before, bacteria are able to take up DNA which is 'loose' in their surroundings, and by encouraging them to do so, or by several other tricks developed for this purpose, genetic engineers can now easily insert their DNA into a bacterium.

So now we are all set up to make genetically engineered insulin. We know how to take human DNA, chop it up with an enzyme, and splice it together with the Lac promoter so that it works properly inside *E. coli* (the promoters controlling human genes work well in the human cells in which those genes are usually active, but are of little use in other cells and no use at all in *E. coli*). We can splice the result into a vector molecule so that it duplicates efficiently as the *E. coli* grows, and generate an *E.coli* containing the engineered DNA. Then we separate that *E.coli* from all the others . . .

But which *E. coli* is it?

Selecting the right clone of bacteria was the other major headache for genetic engineers. We are back to needles in haystacks again, but with an important difference. Now we have the potential for making large amounts of DNA, and for getting that DNA to direct the cell to make a specific protein, neither of which was necessarily true for the genes as they were in the human cells.

One preliminary test is simple. We can test for which bacteria contain a plasmid molecule by including in the plasmid vector the gene for, say, penicillin resistance, and then growing the bacteria we wish to test in the presence of penicillin. Thus any cells which had not received a copy of the vector molecule will die off, leaving only those containing the engineered vector.

A similar process can be used to see if the gene in the engineered E. coli is producing a protein we want. We spread the E. coli out so that they are well separated one from another. Then each E.coli is allowed to grow into a small clone, so that each original E. coli is represented by several million identical descendants, a much easier target for chemical tests. Typically, the test will turn the target bacterium blue or brown, and we can then select that bacterium and discard all the others simply on the basis of their colour. Such tests often use antibodies as testing reagents. We mentioned in Chapter 4 that antibodies are proteins which can latch very specifically on to other molecules as part of the body's defence against invaders. Here we use that specificity to detect our 'new' protein inside a bacterium which is already full of all sorts of other proteins. Of course, this is not easy to do in practice. The right promoter has to be hooked up to the right gene, and both have to be the correct way round (which is easy to ensure if you are dealing with string or girders, but rather hard with molecules). This also assumes that our gene will work properly in E. coli, and that is an assumption which is often untrue because the exact mechanism of how genes are put together, the 'grammar' of genetics, varies quite considerably in such diverse organisms as man and E. coli (we shall be considering some of the differences in Chapter 10). However, it is quite an attainable goal, so much so that the major problem with this sort of approach nowadays is not the production of bacterial clones but the production of antibodies with which to test them!

There are several other methods which can be used to test the bacterial clones in order to isolate the right one. One of the most popular in research laboratories is the technique of hybridization, which we will meet in a slightly different guise in Chapter 9. All these tests are derived from classical biochemistry (in molecular genetics, anything much over a decade old is 'classical'), so once the recombinant DNA techniques were worked out – the cutting and splicing of DNAs – a battery of tests followed quickly. The production of recombinant DNAs is the *sine qua non* of the whole method, and therefore the whole package of techniques is often called 'recombinant DNA technology' to reflect this prominence.

The recombinant DNA technology I have described so far is now relatively easy to apply. What was in the forefront of research only ten years ago is now a technique which a graduate student can feel confident in making the subject for his or her thesis before getting on to something more interesting. The only question remaining, then, is: Which bit of DNA do we want in such large amounts? What DNA should we start with?

It depends on what we want the genes to do. The obvious method is to take DNA from another organism, one which already has the gene we want to engineer. Thus if we want to make a bacterium which contains the human insulin gene, we would start with human DNA. Unfortunately this does not always work, because of the differences in the 'grammar' of the genes to which I alluded above. This will not matter if we just want the DNA for study: indeed, it is by such study that the differences were discovered in the first place. But if we need the DNA in order to produce a protein, then a more 'grammatical' form must be found. Two other approaches might be suitable: making a cDNA or making a synthetic DNA. A cDNA is a DNA copy of an RNA – the name stands for copy-DNA. (It also stands for 'complementary DNA', because the DNA has a base sequence complementary to the RNA.) By using an enzyme to copy an RNA on to a DNA (remember that the RNA was originally copied from a DNA in the first place in the cell) we can get round some of the 'grammar' problems, as these are a feature of the DNA and not of the RNA. However, sometimes there is not even a suitable RNA available, or, if there is, there is no way of testing the bacterial clones we produce to see which one has the right DNA in it. Then the genetic engineer has no choice but to make the DNA from scratch in a test-tube, using chemical reactions to link the bases together one at a time. As genes are rarely less than a couple of hundred bases long, this is quite tedious. Enterprising companies have recently marketed machines which will do all this automatically, making DNA molecules up to 100 bases long in twelve hours or so. We can then use DNA ligase to join up a few such molecules to make our new, entirely artificial gene.

Even when we have the correct DNA, we still have to link it up to suitable vector DNAs, put a promoter at its start if we want to make a protein, insert the DNA into a bacterium and then find which bacterium has received the DNA we want before we get any of the protein product or any of the 'pure gene' we were after. There is obviously quite a lot of work involved and so using genetic engineering to produce a protein can be a very expensive business. This is why genetic engineering of bacteria to make a protein product has so far been applied only to the production of materials that are extremely costly to produce by other methods, thus guaranteeing a good return on research investment. Insulin is an actual example of such applied genetic engineering, which we will meet again in the next chapter.

Bacteria can do many things, but they are not amenable to use in every kind of manufacture. For example, baking and brewing are presently carried out with the assistance of yeast because these organisms are very much better suited to performing these processes than are bacteria. So if the genetic engineer wishes to try to improve methods in these fields he must apply his technology to yeast. Yeasts are single-celled organisms like bacteria, and, like them, tend to be generalists in their approach to life. They also have plasmids which can be used to make vectors for genetic engineering. There are, though, some fairly fundamental differences between how genes are organized in yeasts and in bacteria – yeasts in their genetic 'grammar' bear more similarity to humans than they do to bacteria, so the various controller genes and vectors which work well in E. coli will not be usable in yeast. However, this need not concern us here, as the final message is that we can consider doing the same sort of gene-swopping experiments with yeasts as we have outlined using E. coli. We will mention yeasts again in Chapter 11, and also the possibility of using animal and plant cells in analogous types of experiment.

I have been talking about splicing genes into bacteria with a view to deliberately altering the bacterium's make-up, with only the immediate goal of the genetically engineered bacterium in mind. We try to make a bacterium synthesize a protein it has never made before, and, in the case of biomedical products such as insulin, one which it is unlikely ever to make in nature. This sort of genetic engineering, putting 'foreign' genes into bacteria, is already in very wide use. But looking beyond the immediate product to the longer term, we might reasonably ask: Is it safe?

Today scientists consider recombinant DNA techniques to be quite safe, with the proviso that they are used with the caution all researchers in all fields should apply as second nature. But a decade ago things were very different. In the early 1970s the various techniques we have described came to fruition within the space of a few years, making it seem as if genetic engineering had become a practical reality overnight. No safety testing had been done, as the subject had not existed before. No reasonable predictions could be made on the basis of past experience: in this new technology, there was no past experience. So scientists had to speculate on the outcome of their experiments, and the speculations sometimes got quite out of control.

Three types of catastrophe were envisaged.

All early genetic engineering dealt with bacteria. The bacterium used, a particular 'race' of E. coli, is harmless, but E. coli is a natural inhabitant of our intestines. Human beings, and many other animals, harbour this type of bacterium in their gut, and although usually benign it is not always so. Some types of E. coli can cause infant diarrhoea or food poisoning, both dangerous diseases. By exchanging genes between bacteria, scientists could put the gene for a dangerous protein (for example the proteins which kill lockjaw (tetanus) or botulism victims) into bacteria which flourish in our own gut. The result could be a new disease with the infectivity of a 'tummy bug' but the lethal effects of botulinus toxin poisoning. It could be resistant to antibiotics, too, as many of the plasmids used as vectors for genetic engineering experiments carry the genes which give the cell resistance to penicillin, tetracyclin and sometimes to other antibiotics too.

The protein need not be a recognized poison. Some crude forms of psychological treatment involve injecting huge amounts of insulin into patients, who go into shock, occasionally with beneficial results. The dose must be carefully calculated, as too much can send them into a fatal coma, a risk that also hangs over diabetics who have to inject themselves with insulin. But genetic engineering is proposing to make an *E. coli* specifically designed to manufacture vast amounts of insulin. What if an insulin-producing *E. coli* got into someone's intestines? The dose of insulin produced might be enough to kill, even though in smaller doses it is essential to the body.

Such fears were not eased by the way that plasmids could swap between bacterial cells. Plasmids and viruses share the characteristic that their DNA can be passed from cell to cell. Genetic engineers use this very property to get their recombinant DNA molecules into cells in the first place. But this facility means that, in principle, those same genetically engineered DNAs could be passed on to other bacteria, ones not so benign as those used in the laboratory.

A second group of less well-defined scenarios suggested the possibility of infectious cancer being generated. We mentioned in Chapter 4 that certain genes, the oncogenes, may be an essential part of the mechanism by which cancer starts. If those genes were accidentally or deliberately cloned in a bacterium and the bacterium escaped from the laboratory, the new bacterial strain might cause cancer in everyone it infected. There is still no cure for cancer.

On a more general scale, we must consider genetic engineering in its evolutionary perspective. For over a billion years bacteria and the animals and plants have evolved separately, setting up a barrier which genes were believed in 1975 never to have crossed. By exchanging genes between man and *E. coli*, or between yeasts and plants, that barrier is being breached. Two quite different types of life are being mixed up. Is this really a good idea? (Actually, recombinant DNA techniques themselves have shown this argument to be groundless, by demonstrating that genes *have* passed between species as diverse as plants and bacteria. But that was in the future.)

In 1975 these fears, coupled with the rapid increase in the use of genetic engineering, led to a worldwide ban on using recombinant DNA techniques while scientists tried to sort out the myths from the facts. The most extraordinary thing about the ban was that the scientists imposed it on themselves – the first time that any major branch of science has brought itself to a halt for other than technical reasons. Later on, the scientists were most puzzled when the public objected to their starting up again; politically naive, they had assumed that whatever they passed down as being proven by experiment would be accepted by politicians, who had in fact very different axes to grind. The public remained suspicious that the scientists were doing this because they believed themselves to be the only ones capable of judging their own activity. They could stop research, and they could start it up again, when they chose. As a result this bold experiment in social responsibility generated a lot of ill-feeling within the scientific community and between scientists and legislators, most of which only died down some time after the real facts had emerged.

The fears expressed then were often reasonable at the time (although the then Mayor of Cambridge, Massachusetts, who said that he did not want seven-foot monsters climbing out of the sewers in *his* town, was confusing science and science fiction), but have since been shown to be groundless.

However, it takes some strong evidence to be sure that such a powerful methodology as recombinant DNA is safe. How was this done?

The picture has altered in three ways since those first days. Firstly, scientists took the obvious step of conducting all genetic engineering experiments in closed laboratories, so that none of the genetically engineered bacteria would get out if they did turn out to be dangerous. A standard set of containment conditions were developed, called P0, P1, P2, P3 and P4, based on tried and trusted systems for keeping bacteria under control. These had been developed during more than fifty years of work on highly infectious diseases in hospitals and 'isolation units'. P0 corresponds to normal laboratory procedure. P2 represents locked laboratories with a 'negative pressure' system (a method of air-conditioning which keeps the pressure inside the room a little lower than that outside, so that leaks always leak in, not out), sterilization of all the waste that leaves the room and so on. P4 is equivalent to the facilities used in the biological warfare research laboratories of the 1950s. P4 laboratories are so complex that only a few have been built for civilian work anywhere in the world.

Secondly, scientists changed the bacterial cells in which they were going to place their recombinant DNA. Not all bacteria are equally robust, and scientists can breed unusually weedy ones deliberately, using the science of genetics to select mutant bacteria which would never normally survive in nature. These can then be used for genetic engineering. Some now in use are so feeble that it is hard to get them to grow at all, so in the rough and tumble of the world outside the laboratory these bacteria would last only a few minutes. In fact, all laboratory E. coli are rather enfeebled by their long history in the lab; although this bacterium has been a standard subject for laboratory research since the 1960s, the first case of human disease caused by a laboratory-type E. coli was recorded in May 1981 – and that was not in a genetic engineering laboratory.

Thirdly, in conjunction with feeble bacteria, mutant vector DNA is used. As we mentioned earlier, both plasmids and viruses carry genes which help these DNAs to pass from one cell to the next. In the new, safer, vector DNAs these genes are mutated so that they can only work in the laboratory bacteria (in the case of viruses) or are removed entirely (in the case of plasmids). So the vector DNAs cannot spread to other, tougher, bacteria if they accidentally escape from the laboratory. Theory predicts that when both feeble E. coli and special vectors are used together, the chance that the resulting recombinant DNA can escape from the laboratory is so low that the likelihood of being struck by lightning during an automobile accident is a near certainty in comparison.

But we need rather better assurance than theory that such precautions are going to work. These are all 'human' precautions, and as such can go awry if someone forgets a vital step or misreads a simple instruction. Luckily for commercial genetic engineering, such assurance has come from three quarters.

Firstly, there have been no accidents – not merely no infectious cancer and no new killer plagues, but apparently no illness or environmental disruption of any sort caused by genetic engineering. (It may seem rather irresponsible to go ahead with research and then to say *afterwards* that it was safe because nothing went wrong, but do not forget that research in the early days was all carried out in sealed laboratories under stringent controls: no one was taking chances until they had the facts. Like a potential bomb, genetic engineering was watched closely in the closed laboratories, sealed off from the outside world, until everyone was sure that it was not going to explode.) This is a remarkable record, one which other new industries like the nuclear power or even the microelectronics industry can scarcely claim. Microbiologists, who have dealt with dangerous bacteria for over a century and have developed very effective methods for controlling them, were not very surprised at this. Indeed, they were quite annoyed at all the fuss being made about recombinant DNA techniques in the first place. They had had to contend with *known* dangers, like rabies and smallpox, and the concern over these hypothetical problems seemed to them rather excessive. Experience has proved them right. Thousands of laboratories employing tens of thousands of individuals around the world perform recombinant DNA research, and the record shows that it is safe.

Secondly, as the techniques themselves revealed the details of how genes were put together, it became clear that merely putting the gene for insulin into a bacterium was not going to produce a plague. A wealth of control DNA must be arranged in just the right way before scientists could get the genes to work at all, let alone run away with themselves. The genetic mechanisms became more and more complex the more closely scientists looked at them, and the chance of making them do something by chance alone became correspondingly remote, like the chance of making a Swiss watch by throwing a lot of small gears into a metal box and shaking them. Furthermore, the 'grammatical' differences between the genes of man and *E. coli* mean that even if we did let an *E. coli* out of the laboratory carrying an oncogene, it could do nothing with it: the message carried by the gene would be gibberish to the bacterium that was carrying it. (We shall mention these differences again in Chapter 10.)

A curious piece of knowledge was turned up by the new

science: the barriers between animals and bacteria which seemed so impenetrable are actually not so solid after all. Several bacteria have been found which seem to have animal genes in them, and a whole group of plants have a type of haemoglobin in parts of their root system. We mentioned haemoglobin in Chapter 5: it is the red pigment that carries oxygen in the blood of animals. No one is sure, but there is a strong suspicion that the plant got the gene somehow from an animal. So the almost philosophical worries about swapping genes between species like man and *E. coli* have receded as it has become clear that such things can happen in nature anyway, without disastrous consequences.

But perhaps most reassuringly, studies which looked directly at what happened to *E. coli* when released into the outside world came up with some surprising results. Volunteers were fed large numbers of *E. coli* – none of the bacteria survived to emerge in the volunteers' faeces. *E. coli* were deliberately tipped on to laboratory benches – they all died in under thirty minutes under artificial lights and in a few minutes in sunlight. And the more genetic engineering a scientist did on a bacterium, the feebler it became. Almost any change they made in the cell's genes was a change for the worse – not from the scientists' point of view, of course, but from the cell's. It is a step downhill, away from the evolved perfection that a billion years of evolution has given to today's living things. A genetically engineered bacterium is always less fit to survive than its unengineered siblings in the struggle for existence. It cannot cope so well because it is carrying all those extra genes around, and nature rewards these burdened bacteria with the cruel justice of survival of the fittest. When faced with competition from all the other bacteria in the world, only the unaltered bacterium can survive outside the scientist's flask. Even in the laboratory, bacteria show an annoying tendency to dispose of their genetically engineered genes. They have to be constantly checked to ensure that they have not thrown a bit of their plasmid away, or thrown the whole plasmid out of the cell. And a plasmid making a protein – any protein – can be a disaster for the cell. The extra energy needed to make that protein leaves the cell panting along behind in the race for survival.

Perhaps this is not surprising, for we are not 'improving' nature but merely using her facilities for our own convenience. Something has to pay for that convenience, and that something is the bacterium we engineer.

Of course, this does not preclude the possibility that an irresponsible person might go out of their way to make an unsafe bacterium, to generate a new Black Death, using the techniques of genetic engineering. It would be hard to do, considering all the barriers in the way, but human ingenuity seems limitless. Could every one of man's and nature's safeguards be sidestepped?

That possibility remains in the future: we will be talking about it in Chapter 13. The present has arrived in numerous more rational areas of genetic engineering. The techniques we have surveyed, the cutting and rejoining of DNA, making new plasmids containing new genes and their controller genes, have been used in industry and academe for over a decade. In the next chapter we shall see how the ivory-tower science of recombinant DNA came down to earth as the commercial technology called 'genetic engineering'.

Chapter 8
The First Applications

In 1979 a small research company called Genentech put its shares on the open market. Their value doubled in a few hours of frenzied trading. Among the future products of their new 'recombinant DNA technology' they listed insulin, interferon, new vaccines and hormones, and everyone wanted to be in on the ground floor of the new science.

Commercial genetic engineering had arrived.

Why did recombinant DNA technology open with such spectacular hopes in the late 1970s, and then suffer such a dramatic slump of popular interest? The seeds for both boom and bust were laid in the preceding years. The boom was based on three things. The researchers needed knowledge about genes and how they worked if they were to manipulate them successfully – not the indirect knowledge that was the legacy of a hundred years of genetics, but an exact description of DNA as a store for genetic information. They needed the enzymatic tools to manipulate the DNA once they had begun to understand how it worked. And they needed some problems to attack once they had amassed all that technology. We have seen how the problems we might wish to address are many and varied, from making enzymes to curing cancer. In the mid 1970s the techniques needed to do something about them came to fruition and provided the tools to attack some of the simpler projects.

These projects fall into three types. Research continues on the structure of genes, with feedback from recombinant DNA technology generating more precise information, which in turn is used to generate more powerful techniques to gain yet more information. Indeed, this has started a chain-reaction of information-gathering and a consequent information explosion which we shall meet again in Chapter 10. The other two applications are, however, more directly practical. Both have taken the simplest thing a gene can do – which is to make a protein – and have used it to make a saleable product. These applications are the production of proteins which are intrinsically useful to us, like insulin, and the production of enzymes to perform an industrial process.

We recall from Chapter 6 that one sort of gene is a gene whose information content instructs the cell how to make a protein – it is the gene 'for' that protein. Now, there are many proteins which would be very useful to have in larger amounts than were available in 1970. Enzymes can be used as catalysts in industrial processes, doing quickly and precisely what is now done less efficiently in chemical reactors at high temperatures and pressures. Other proteins, like hormones, neurotransmitters or vaccines, could be valuable for medical use. All these things are made in nature, of course: for example, all mammals make insulin. But they do not make very much of it. When Frederick Sanger determined the sequence of amino acids in the insulin molecule he only used a few grams of protein – not a lot, by everyday standards. But it still took about 125 cows' pancreases (the usual source of insulin) to produce this much. Diabetics, who use insulin to control their disease, used the insulin from sheep and pigs to inject instead of human insulin, because although it is slightly different it is in rather better supply. However, the number of diabetics is increasing at about 6 per cent per annum in the English-speaking world, and in the fairly near future all the pancreases of all the pigs and sheep killed for food in the West will no longer be enough to provide insulin for this growing band. So it would be very useful, and profitable, to have a new and plentiful supply of this protein, and preferably the human protein as there are fewer side-effects associated with its use.

Other medical proteins are obvious potential targets. Growth hormone can be prepared from human cadavers, but it takes the hormone from seventy cadavers to provide treatment for just one year for a patient deficient in this hormone. Again, a more plentiful supply is urgently needed.

The path which industry followed in making these proteins was similar to that which we have followed in the last few chapters in unravelling the technology of genetic engineering. The earliest results used only the genetics of the plasmids, without any of the sophisticated DNA cutting-and-splicing operations which characterize recombinant DNA research. Usually plasmids are not just spare genes along for the ride; they often contain genes for functions vital to the bacterium that holds them. In particular, in some soil-living bacteria of the group *Pseudomonas*, plasmids carry the genes for breaking down the molecules of complex hydrocarbons like those found in oil, producing chemicals useful to the bacterium as the product. This can be a very complex process, and not one which a small number of enzymes can perform on their own. Each plasmid consequently carries the genes for only a few of the enzymes necessary for the complete degradation of oil, making up a specialist package to break down oils, petroleum, tar, and so on. None of these plasmids is very efficient in chemical terms. The bacteria which contain them cannot break down a lot of oil quickly, as breaking down oil is only a secondary activity for *Pseudomonas*. Oil is not a common substance in the soil where these bacteria evolved, so they have plasmids which produce enzymes to cope with only small amounts of it.

Then came man.

One of the more unpleasant aspects of this century's thirst for oil is the accidental – or occasionally deliberate – release of hundreds of tonnes of crude oil a year into the sea. It takes the bacteria living in the sea years to break down such slicks into soluble chemicals, and so we speed the process up, usually by spraying the oil with detergents and skimming off the resultant scum. In 1970 the General Electric Corporation thought of a better way. Could those *Pseudomonas* bacteria be encouraged to dissolve the oil for them, growing on it while breaking up the slick? Then the bacteria would be eaten by plankton, which would in turn be eaten by fish, so that in a few weeks nothing would remain.

General Electric isolated the plasmids which made enzymes capable of breaking up the various components of oil, and planned to place all their genetic information in one bacterium. This bacterium would then be able to attack the components of oil – it would be an 'oil-eater'. In 1972 Ananda Chakrabarty filed a patent for the bacterium on behalf of General Electric.

There were two hitches in this genetic engineering, both deriving from the use of unaltered plasmids as the carriers of the genes in which we are interested. As we mentioned in the last chapter, swapping the plasmids around will tend to produce a bacterium which is less healthy than its 'wild' counterparts. If it accidentally loses one of the new plasmids, it will find that it has an advantage over the other engineered bacteria and so will eventually outgrow them. This is an example of natural selection at work, and it will continue until the bacterium is as fit for its environment as it can be – this usually means that it has lost all the genes we have tried to give it. Thus General Electric needed a method of stabilizing the genes they were putting into the *Pseudomonas* so that they were very difficult to lose by accident.

The other problem was that the genes found in nature are inefficient at producing the oil-dissolving enzymes. Before General Electric's 'oil-eater' could be a commercial proposition, its efficiency had to be improved.

The same difficulties have plagued other attempts to use bacteria to break down waste materials. Several companies now market bacterial preparations for disposing of noxious chemicals. Sun Oil used bacteriology to clean up an underground oil spill that was threatening ground water supplies in Pennsylvania, and Flow Laboratories, the Polybac Corporation and Sybron Biochemicals all market bacterial cultures to clean heavy oil- or grease-like blockages in pipes. However, these are specialist applications, because usually the bacteria are no more efficient cleaners than detergents and hot water.

The problem lies in the efficiency with which the bacteria make enzymes to break down the waste material, and the efficiency of the enzymes themselves. The former is determined by 'controller genes', the latter by the structure of the enzyme itself. That structure is, of course, determined by the gene for the enzyme. Either way, genetic engineering holds out hope that the efficiency of these bacteria may be improved if we can find out enough about the genes concerned to be sure of how to manipulate them.

The question of efficiency also dogged the workers at British Petroleum. They had found another soil bacterium which could convert methane to methanol. Methane is the main constituent of natural gas, and is a very plentiful raw material. Methanol is the starting material for a wide range of industrial processes such as the manufacture of plastics and drugs, as well as being a useful solvent in its own right. Just one enzyme is involved in this reaction and, unlike most enzymes, it is not terribly fussy about what molecules it attacks. It catalyses similar chemical reactions on a number of other, related chemicals, including the synthesis of propylene oxide from propylene. Propylene is a cheap product of oil refining, while propylene oxide is another starting material for the chemical industry, but one which is presently rather hard to make and hence rather expensive. Naturally, BP was interested to see if the bacterium could carry out these conversions on a commercially useful scale.

Like the 'oil-eater', the bacterium given to the researchers was inefficient by industrial standards. BP set out to increase the amount of the enzyme responsible for these chemical reactions by using the gene-splicing techniques we have discussed. They tried putting the gene into plasmids which are present in large numbers in cells (so that there would be more genes) and putting the gene next to a very efficient promoter (so that each gene was more active). As this is an ongoing project, BP have not told the outside world too much about what success it has had. However, the prospects seem quite good, as Standard Oil started to investigate this same technology in 1980.

These projects are attractive to industrial managers because they are likely to be profitable in the long run, but they are not so exciting to the molecular biologists because they are not major departures from the techniques of traditional genetics. At least to start with, the projects used normal breeding techniques coupled with a knowledge of the special characteristics of plasmids to generate bacteria carrying combinations of genes not previously seen in nature. But this use of plasmids pointed to a new approach, the use of the DNA cutting-and-splicing technology we examined in the last chapter to make recombinant DNA molecules out of DNA from very diverse sources. In 1970 a research company called Cetus was started near San Francisco, California, which was to specialize in using these techniques to produce not academic results, which had been their sole 'product' to date, but commercial products. In 1974 it was joined by another San Francisco company called Genentech, again hoping to apply the new techniques to the old problem of making money.

The key word for several years was 'hoping'. These technologies were so new that in 1974 they were still laboratory curiosities, with enormous potential but no track record. The companies accumulated capital, and critics who claimed that they were using technical double-talk to take people for a ride. In reply they said that the science they had invested in was only four years old, and with a few more years, and dollars . . .

And in 1977 the sceptics were confounded. Paul Berg, a professor at Stanford University who was one of the founders of the new technology, although not of the companies that exploited it, was called before a Senate sub-committee investigating whether this new and (as it was then believed to be) potentially dangerous technology was actually of any use. The high-point of the US government's regulation of recombinant DNA technology came in 1977. The debate about the technique's safety was at its height and not a single useful product had emerged – people were calling for a complete ban. So Paul Berg was called as a notable proponent of the new methods, and to everyone's amazement said that his colleagues at Genentech had used recombinant DNA techniques to make a bacterium which could synthesize the human hormone somatostatin. Actually, the announcement was rather premature, as Genentech had yet to get the hormone out of the bacterium in other than minuscule amounts. But the news was out, and Genentech, taking the aggressive publicity policy it would follow for the next decade, made the most of the announcement to garner more support.

Not that it needed much help. The conference at the Asilomar Center which had first raised the issue of the safety of these techniques had taken place less than five years before, and the enzymes which were essential for the cutting and joining of DNA had only been discovered a few years before that. If the new science could make its first commercial product in under a decade of its creation, what might another ten years bring? Enthusiasm among scientists and investors soared – the new age had dawned. Genentech was just riding an irresistible wave of over-enthusiasm.

Genentech performed this engineering operation exactly as we have described in previous chapters, with a few technical frills. Somatostatin is a short peptide hormone concerned with the control of how fast we grow – we mentioned it in Chapter 4. There is no single gene for somatostatin, as it is made by cutting up a much larger protein in the cells of the brain. So Keiichi Itakura and his team at Genentech made the gene chemically, linking the bases together to make the two DNA chains necessary to end up with a double helix. This double helix they then linked up to the promoter from the Lac genes, which ensured that the synthetic DNA would be used by an E. coli to make a protein. They then put the whole construction into a plasmid, and then this new plasmid into E. coli cells. The E. coli cells began making somatostatin, and far more of it than the same number of human cells could produce.

Unfortunately, there was a hitch – one of a characteristic set of hitches that were to plague genetic engineering for some time. The E. coli did not like being full of this foreign protein, so it made another enzyme to break it down again into its component amino acids. Thus the engineered bacterium was making somatostatin at a terrific rate but destroying it just as quickly, so that none escaped for the scientists to collect.

Nowadays there are mutant E. coli which are less liable to break down foreign proteins in this way, although they still do so. However, in 1977 these mutants had not been discovered, so Genentech solved the problem in another way. They joined the synthetic somatostatin gene on to the end of the Z gene in the Lac cluster. In this new recombinant bacterium the E. coli read the gene for somatostatin as the end part of the Z gene. Consequently it produced a protein from this gene which had the amino acids of the somatostatin protein tacked on to the front of the Z-gene protein. The product, a 'fused' result of a normal *E. coli* protein and a foreign protein, was recognized by the *E. coli* cell as a normal component of its cell, and so it did not destroy it. An added bonus is that this procedure ensured that the new protein had a signal peptide on it, so that the cell automatically secreted the new protein in the liquid in which the *E. coli* were growing. Thus the Genentech scientists could get their 'fused' protein without having to break open the *E. coli* cells and letting all the other *E. coli* proteins out into the mix. A single chemical reaction could then chop the somatostatin part of the 'fused' protein away from the rest, giving them the hormone they were after.

Somatostatin was but a foretaste. In mid 1978 the Genentech team announced that they had joined a gene for insulin up to the Lac promoter and a plasmid vector, put the result into *E. coli* and created a bacterium which made human insulin as directed by the new gene. (They did not use the gene from human DNA as it would not have functioned properly in *E. coli*, owing to the 'grammar' differences we mentioned before. Instead they used a cDNA (see page 83).) Even the insulin which was usually used for medical treatment, and which was made from sheep and pigs, was expensive. Some patients react badly to this animal version, and can only be effectively treated with human insulin. Doctors would dearly have liked to have a supply of human insulin that was cheap enough to use on all the diabetics who needed it, and this was the jewel which Genentech offered them in 1978. It is a gem of no mean size either, with two million diabetics in the English-speaking world alone, a substantial fraction of whom need daily injections of insulin just to keep alive. Interestingly, while somatostatin had been the product of collaborative research between Genentech and the University of California at San Francisco (UCSF), insulin was largely a product of Genentech alone, sponsored by the pharmaceutical giant Eli Lilly. Genetic engineering was visibly moving away from academia and into the market-place.

While Genentech grabbed the headlines in the United States, a group headed by Professor Charles Weissman of the University of Zurich was trying to start a similar venture in Europe. Their company was to be called Biogen and the product they first made their target was interferon. By now the rumours of the wonders of recombinant DNA had crossed the Atlantic, and Weissman thought that money would probably be thrown at a European entry on the ground floor of this new wonder industry. They tried first to obtain finance in Britain, believing that the strength of the British universities in pure research and the government's policy of supporting the basis of the industries of the future would make their project an ideal investment. But they were turned down by several banks, and even by the National Enterprise Board (a government-created body set up specifically to provide venture capital for new technological endeavours), who all thought their project too risky. However, the notoriously conservative bankers of Zurich disagreed, and Biogen set up its laboratories in Geneva (it is notable that genetic engineering companies, like the semi-conductor research firms, set up shop in some of the world's nicest places).

In 1980 they succeeded in putting a cDNA gene for interferon into E. coli. The announcement was made by Schering Plough, another pharmaceutical multinational which had seen the future in this strange new technology and had invested in it. The E. coli made about two molecules of interferon per cell per hour, or about one million per second in a pint of bacterial culture. This is an unimpressive figure, but interferon is so incredibly potent that very little is needed for medical treatments. Still, Biogen had to improve this by many thousandfold before they could make enough to be commercially useful and corner the market in a drug which had been claimed to be able to cure every disease caused by a virus.

Schering Plough also hinted that interferon could be a general cure for cancer. This was not a wise move, as the scientists working in cancer research believed this to be only an outside chance. As more interferon has become available for research their doubts have been justified, because interferon on its own has proved to be almost completely ineffective against nearly all cancers. However, it is characteristic of those early days of euphoria that even a slim chance of benefit was amplified into a certainty.

It was also inevitable that in the following years, when these unrealistic hopes were dashed, the public should see genetic engineering as having failed to live up to its early promise. What the researchers *really* wanted from the genetically engineered interferon was enough of it to find out just what it did do, as it was such a rare protein that only a few preliminary experiments had been performed on it. So the researchers basked in the unaccustomed publicity, and hoped the reaction would not be too destructive.

Not surprisingly, the larger-scale trials showed that interferon was far from being the wonder drug made out by some. It is effective against some rare cancers, but most are unaffected. It is an effective treatment for many diseases caused by viruses, and can even cure colds if given in large enough doses. However, at hundreds of dollars a shot, most people would prefer to suffer the sniffles. And there are side-effects, most of which are still only partly identified.

But that was all for the future. In the late 1970s scientists, media pundits and investors were so amazed that it worked at all that they could think of no reason why the techniques of recombinant DNA should not immediately revolutionize bio-medicine. The hard part, surely, was cloning the gene: the rest was standard . . . wasn't it?

No, it was not. There was to be a long haul between having a gene spliced into a plasmid and having a protein sealed in a syringe. Eli Lilly, a veteran in the mass-production of drugs, had Genentech's insulin-producing *E. coli* in 1978, but only set up a test-plant producing insulin in 1980, and marketed the insulin for human use in 1985. This is a snail's pace compared with the rate of recombinant DNA research: in the same period, several entirely new classes of regulatory genes, the novel 'punctuation' of mammalian genes, genes that can double as viruses, onco-genes, and genetically engineered plants and animals, all poured out of the research laboratories. But Eli Lilly was in fact working at an unusually fast pace for the development of a new drug or food. In 1979 the gene for the human growth hormone was cloned, but clinical tests to see whether it could be used to treat patients deficient in growth hormone could not get under way until 1981, and a medical product will probably be available at about the same time as this book is published. A genetically engineered drug has to go through all the long tests that any other drug has to pass before being used on the public. This is the reason that the genetic engineering 'revolution' of the 1970s is only now beginning to affect our lives. (However, it is not the reason why the grandiose hopes of the 1970s have failed to materialize: most were shown to be empty dreams long before the scientists involved started the long road to commercial exploitation.)

There are two main reasons for this delay. Firstly, the law rightly requires rigorous testing of any new medication before it can be released. Such tests take a long time, as the product has to be shown not only to be safe for short-term use but also to be free from long-term, cumulative side-effects. This was one problem which turned up with interferon, as this protein was found to affect the long-term growth of isolated human cells and so might adversely affect the growth of cells in patients undergoing interferon treatment. This potential problem has not been resolved, which is one reason why interferon is still a research tool and not yet a clinical product.

This delay was expected, as the laws governing the release of new drugs are well known to the pharmaceutical companies sponsoring much of the work. But as well as the expected hold-ups there were some unexpected hitches which cast doubt on the whole notion of using a bacterium to make a protein, and these were evident from the earliest days.

We have already seen one of them. *E. coli* is eager to dispose of unwanted proteins within its cells by breaking them down into their component amino acids. Getting round this problem can be achieved by producing a 'fused' protein, such as the one described above. However, processing such proteins is another costly step, and uses an extremely poisonous chemical which must be carefully removed from the product before it can be of use to man.

The protein product is not always exactly what we want, even when we get it. In broad outline, the way in which the genes of *E. coli* work is the same as the way in which our own genes work:

the order of the bases in the DNA is 'transcribed' into an order of bases in RNA, and that is 'translated' to an order of amino acids in proteins. However, the process in animals does differ somewhat in detail, especially in what happens after the 'translation' step.

After translation (i.e. the production of a protein by stringing amino acids together in the right order) comes a series of processes known collectively as 'post-translational modification', a catch-all phrase for a variety of subtle chemical changes which the cell can perform on newly-made proteins. They might link a sugar molecule on here, or clip off a couple of amino acids there, or subtly modify that amino acid three links from the end. These modifications are much more parochial than the great generalities of the genetic code – *E. coli* and humans, separated by at least a billion years of evolution, still use the same genetic code but modify their proteins quite differently during post-translational modification. So there was the question of whether the modifications used by *E. coli* would produce proteins of any use to man. This question arose in the work on interferon – luckily it turned out that the post-translational modifications were not essential in this case. Since then, scientists have tended to avoid the problem by cloning the genes for proteins which are not modified, or only slightly modified, by human cells. But there remains the spectre that one day, after an enormous research effort, the gene for a potential money-spinning new product will be cloned only to be found to be useless because of these differences in post-translational modification.

Thirdly, we have largely ignored the rest of *E. coli* so far, treating it as a box into which we have put a plasmid. But it is really a complex cell in its own right, and is full of DNA, proteins and other molecules which might contaminate our genetically engineered product. Obviously, these are elements which we do not want to inject into a patient together with our interferon or growth hormone, but purifying the product we do want from these unwanted molecules can be quite hard. It can add months or years to a research project, and thus can also make the finished product more expensive.

One last problem is particularly vexatious to scientists trying to produce a genetically engineered product in bulk rather than in the tiny amounts typically produced in laboratory experiments. We mentioned in Chapter 7 that E. coli containing plasmids tend to throw these plasmids away unless they are essential for the survival of the cell. This is good news from the safety point of view because it means that it is unlikely a genetically engineered E. coli will survive long in the outside world in its engineered form, but it is a problem if we want to grow that bacterium in the laboratory and make sure that it retains its plasmid. We can get round this by making sure the plasmid is essential for the bacterium's survival. For example, we could include the gene which gives E. coli an enzyme to break down penicillin. If we grew E. coli in the presence of penicillin, then any bacterium which lost its plasmid would no longer be able to make the penicillin-destroying enzyme, and so would fall victim to the antibiotic. Thus all surviving E. coli must have the plasmid in them — we have made possession of the plasmid essential. This sounds like a straightforward solution, and indeed it is quite easy to do. It is not useful, however, as a commercial solution. Penicillin is expensive, and it would be ruinous to use hundreds of tonnes of it to run a chemical engineering plant exploiting a genetically engineered E. coli. So what is quite cheap to do in a laboratory, where E. coli is rarely produced in volumes of more than a few litres at a time, becomes hopeless when we plan to 'scale up' to thousands of litres.

All the technical 'fixes' we have mentioned so far have significant costs. For rare biomedical products this is not a financial problem, as the return on even a few grams of interferon or growth hormone is enormous and can cover the most sophisticated research and development costs. But for less glamorous projects it can be a financial block to a technically tractable problem.

So far we have mentioned only four products of genetic engineering, but there are far more. They fall into three classes. The use of genetic engineering on plants and animals we will leave to later chapters, as this area has its own unique problems and excitements. The other two areas are the production of other medically useful proteins and manufacture of enzymes. Both are high on the list of things we would like to do with genetic engineering.

Somatostatin, insulin, interferon and growth hormone are all 'biomedical' products, the result of a collaboration between modern biology and medicine. It is with good reason that they were the first type of product to be produced by genetic engineering. They are very potent, with only tiny amounts being needed to give a powerful effect, and therefore the genetic engineer could aim to produce the protein in grams rather than tonnes. Before gene cloning they were rare because they are hard to produce – for example, it would take about 500,000 sheeps' brains (the usual source) to make as much somatostatin as two gallons of genetically engineered *E. coli*. Consequently these substances are expensive, holding out hope of a good return on research investment.

There are many other products with the same qualities, and all are currently the target of gene-cloning research in academe or industry. Among them are a host of 'regulatory factors', proteins which signal to the body's cells how they ought to be acting. These can control growth (like the growth factors), the immune system (such as the interleukins), or the function of other organs like the brain (as do the endorphins). The genes for these proteins have all been cloned, and will soon be used to provide a plentiful supply of those proteins for research and, ultimately, for medical use. Among the hormones which are being pursued are leutenizing hormone releasing factor (which regulates fertility), gastric inhibitory peptide (which controls acid production in the stomach and hence could be used for ulcer treatment), parathyroid hormone (which controls bone development, among other things), tumour necrosis factor (which may help the immune system to fight cancers), and a host of others. These diverse proteins all have one thing in common: they are the body's own method of regulating a particular function, and so are powerful tools for altering that function if it goes wrong through disease or injury. Doctors would like to have them in amounts large enough to be used to alter their patients' metabolic processes, and so the genetic engineers try to provide the proteins in such amounts.

Vaccines are also potentially very rewarding products of genetic engineering. A vaccine is something which will cause the body to make antibodies against a potential disease-causing agent. Usually a vaccine against a virus or a bacterium consists of that same virus or bacterium, which has been killed. The lymphocyte cells of the immune system react to the dead virus just as they would to the live one and so make antibodies against it, but the virus does not cause disease while this is going on. The problem with this approach is that the virus has to be one hundred per cent dead, as otherwise we are simply giving the patient the very disease we wish to prevent. It is likely that the outbreak of foot-and-mouth disease in France and southern England in 1980 was due to a batch of vaccine which still had a few intact viruses in it.

A relatively simple way to get round this problem is to use viruses which have been altered so that, while they can still multiply, they cannot cause disease. This has been attainable by ordinary genetic methods, but genetic engineering can offer a quicker and surer route to the same goal. We can use the restriction enzymes and DNA ligase with precision to cut out a crucial gene from the virus, and then rejoin the rest together to make a 'defective virus'. If this is done properly, the resulting virus can be grown in the test-tube quite easily as the scientist can compensate for the lack of that gene, but it cannot grow in the body where it could cause disease. This is a fairly simple type of genetic engineering to perform, as we are only trying to stop a gene from working and not attempting to make one work. Consequently several commercial projects using this sort of genetic engineering to create a 'crippled' virus for use as a vaccine are being pursued at the moment. One such virus, a vaccine against pseudorabies (a disease fatal in pigs although it does not affect humans), has even been tested by its producers, Biologics. It seems to work very well. However, not everyone is pleased with the result, as it appears that Biologics were rather lax about obtaining permission before they actually went ahead with tests. There was no technical reason why they should not test, they said, as the virus was effectively just a safer version of a vaccine already in use, and they had tested both the old and the new virus exhaustively. But whether they were as enthusiastic about reporting their tests is another matter.

Another problem, which genetic engineering a virus to make it less potent has been unable to overcome, is that some viruses and bacteria change their outer proteins every now and again; if this happens, all the old vaccines and all the antibodies which people have in their bloodstream are suddenly out of date, and not effective against the new virus proteins. Of course, not all the proteins in the virus change, only the outside layers – the inside sections are important to the functioning of the virus particle and so have to stay the same. When we change to go out of doors we only put on an overcoat or take an umbrella to protect ourselves against the weather – we do not have a liver transplant at the same time. It is our overcoat, not our liver, that people see when they meet us in the street, and it is the proteins on the outside of viruses or bacteria that lymphocytes detect when they meet them in the bloodstream. So they are the ones that antibodies will latch on to. So, although they are basically the same, the 'new' viruses with their 'new' coats look completely different to the immune system.

Genetic engineering can get round this problem, too. We can make an *E. coli* which will manufacture one of the proteins from a virus, and we can use that protein as a vaccine. The body's lymphocytes will be just as able to recognize it as an invader as they would be able to detect the whole virus, and there is no risk of any intact virus in the protein preparation. If we used a protein which the virus could not alter, then even if the virus did change its outer-coat proteins as part of a coat-changing strategy the immunity given by the synthetic vaccine would not be removed.

That is the idea. In practice it turns out that not all proteins are suitable for making vaccines, as they are only 'seen' very poorly by the lymphocytes. And some of the proteins in a virus may never appear in the blood at all in the course of a normal infection. The hepatitis-B virus, for example, spends most of its life hidden inside liver cells, and the proteins it is made of in those cells are completely different from those which comprise it when it is in the bloodstream. However, lymphocytes can only detect it and antibodies can only latch on to it when it is in the

107

blood. Therefore any vaccine has to be made from the blood form; otherwise the antibodies will let the virus pass by when it gets into the blood and when it arrives in the liver cells will be all geared up to attack it, but then will be unable to get at it. So before we even start the tricky business of cloning the gene for the viral protein, making an E. coli synthesize that protein and then extracting it in large enough amounts and in pure enough form to be a useful commercial medicine, we have to find out which of the dozens of proteins or fractions of proteins which a virus can contain is the best one to use.

Such obstacles might make the production of vaccines by this second method seem too difficult to be worth doing, but that is far from the case. Vaccines against such diverse diseases as bacterial infection of the bladder, influenza and malaria are being researched in laboratories around the world, as are a wide range of veterinary vaccines.

Although dozens of such projects to make biomedical products have been started, no one can tell how many will survive to completion. This is because, as well as the technical problems mentioned above, there is the additional difficulty that no one is really sure what many of these genetically engineered proteins will do. We touched on this when talking about interferon (page 100) — the interleukins are another example. These proteins, which control the cells of the immune system, are a recent discovery; until the advent of recombinant DNA technology there were such small amounts of them available for experimental tests that it was often difficult to see if a 'new' interleukin was actually the same as any of the ones already identified. But scientists know in general terms that these proteins are important in controlling the immune system. That makes them interesting for commercial genetic engineers as they might be useful for boosting the power of that system when we are under attack from disease. However, until we know exactly what they do, and what side-effects they might create, their potential as medical products remains unknown.

A curious product of the greater knowledge brought by genetic engineering techniques is the realization that things are often far more complicated than they seem (a frequent discovery in biology!). For example, there are three types of interferon, called α, β and γ. There is one gene for the γ type, as you might expect, but dozens of genes for the a type. No one knows why. The three types are produced by different types of cells, but they do not act separately. Instead their action seems to combine synergistically – one dose of α and one of γ might have the same effect as three doses of α on its own. Exactly how this combination effect works depends on how you measure it. So what looked like a simple project to clone interferon for use in treating disease has turned out to be a mass of complications with far fewer useful results than had been thought. Genentech, Schering Plough and others are currently looking into which mixtures of different interferons are most effective for treating various diseases.

Why this possibility for confusion was not emphasized from the start is probably most charitably explained as an excess of enthusiasm on the part of those supporting the new technology in the late 1970s and early 1980s. Genetic engineering is not unique in stumbling over the complexity of nature. For decades biologists have been used to the idea that the more closely you look at any living thing, the more complicated it seems to be – this is why doctors take a long time to train, as they have to be able to look in great detail at one of the most complex of living systems, a human being, and make sense of the resulting confusion. Cloning the gene for insulin to produce human insulin for medical use was hard, but there was a wealth of background knowledge about what insulin does and how it does it – the molecular geneticists were not in for many surprises with this well-characterized project. However, to a professional biologist the idea that you could just clone interferon and inject it into humans when no one was sure what interferon did in the test-tube, let alone in an animal, must have seemed fascinatingly naive. After a decade of increasingly complicated research, it appears that way to the 'gene-cloners' too.

Because of the potential for good profits, these biomedical products have been the earliest, the most heavily funded and the most advertised of the products of recombinant DNA technology. Indeed, media enthusiasm for the early result of cloning the genes for insulin and interferon was part of the driving force behind the early stock-market boom in genetic engineering. Since then, problems have arisen and the over-confidence has evaporated, sometimes being replaced by an equally unrealistic sense of disillusionment. In reality, though, the biomedical side of the technology has performed well without producing miracles. Recombinant DNA molecules have been constructed and used to direct E. coli to make new proteins, just as promised. A few of these products are now marketable, and are proving to be very profitable. Others have fallen by the wayside as expensive mistakes.

The other area of application of recombinant DNA has been less spectacular. With a few exceptions, the genetic engineering of bacteria and yeasts to produce enzymes, chemicals and bulk protein has not had the media attention of the biomedical products, but rather has been the subject of steady research and moderate optimism since the earliest days. One reason for this is that there have not been spectacular successes in the sense that cloning the gene for insulin was spectacular. Another is that this side of the research is carried out either in the laboratories of large food or chemical companies or by specialist companies wholly financed by such laboratories. Researchers working in them do not need media attention to attract finance, and the parent companies would rather their competitors did not know what they were doing. Thus such work, while not being exactly secret, is often confidential.

Tate and Lyle were among the first to use enzymes, in sugar processing. As well as the familiar bags of granulated sugar destined for domestic use, Tate and Lyle produce a range of specialist sugar products for the food industry. While conventional techniques are quite adequate for the production of these sugars, recombinant DNA could provide enzymes to make some products more cheaply. So the technology is applied to the unspectacular but very profitable task of reducing overheads in an existing process.

Tate and Lyle are not alone in applying molecular biology to food production. A number of products are potential targets for genetic engineers, who would seek to alter the amounts of enzymes in micro-organisms which are used to make food additives commercially. The altered micro-organisms would, it is hoped, make their particular product faster, or more efficiently, or both. These are some of the present targets: glutamic acid, about 100,000 tonnes of which are made every year by fermentation – most of this output goes into making monosodium glutamate, a condiment used extensively in the Far East; the amino acid lysine, about 80 per cent of which is made by fermentation (i.e. by the action of micro-organisms such as yeast or bacteria – the fermentation of grape juice to make wine is an example of such a process); lysine, used as a food additive in cattle feed to supplement poor protein quality (some humans use it too, in the mistaken belief that their mixed diet gives them no better food value than dried grass); and citric acid, made commercially by a fungus and used as an acidifier in many foods, of which the most obvious are the 'citrus' drinks. All these, and other chemicals, are made by micro-organisms using the organisms' own enzymes, and so could be targets for recombinant DNA research aimed at altering the controller genes to make those organisms produce more of the enzyme, or altering the genes for the enzymes themselves to make the latter more effective. Make more or better enzymes, and the same amount of micro-organism consuming the same amount of food and energy gives more of your product.

This is potentially true of such relatively simple chemicals, but is doubly so for more complex ones – for instance, those used in food flavourings and colourings. These are often quite difficult to make using conventional chemistry, and are usually synthesized from natural products with similar molecular structures. Here the genetic engineer could either aim at producing the natural starting material more cheaply, or even at making the bacterium convert this starting material into the final product as well.

One unlooked-for advantage of this could be a reduction in the number of 'unnatural' chemical additives in food. At the moment the products of the chemical industry are often used to flavour food because an equivalent natural material would be too expensive to use. If genetically engineered bacteria were to make the natural product more cheaply, there would be an economic advantage as well as a medical one.

However, the food business is conservative, and so using something as novel as genetic engineering to produce our daily bread might meet with some marketing problems. The same safety considerations apply here as to biomedical products, of course, and similar technical hitches can arise; although some of these potential problems disappear if we are considering using the chemical product for agricultural or veterinary food and not for human consumption.

Unlike biomedical research, there are far stricter price limits on novel processes in the food and chemicals industry. We mentioned earlier in this chapter the conversion of propylene into propylene oxide, using an enzyme. But I only hinted then that it was not worth doing on a large scale. The cost of the raw materials and the bacterial culture apparatus is such that the end-product would be more expensive than if it had been made by a conventional chemical plant. It takes a lot of complex machinery to keep hundreds of thousands of gallons of bacterial culture medium at the right temperature and properly aerated, well-stirred and circulated, and to remove the product at an efficient speed. So unless using genetically engineered bacteria is a real advance over the more usual way of doing things, using recombinant DNA will be a technically sophisticated luxury that most companies cannot afford.

This is why half-way house operations, such as getting a bacterium to make an enzyme and then using that enzyme just like any other chemical, are sometimes favoured — the conditions for *growth* of the bacterium do not have to be those which are ideal for *production* of the final product. The two processes have been separated. The first-ever success in commercial genetic engineering was just such a project. An *E. coli* was created which contained a gene for the enzyme DNA ligase (which we met on page 79 as the enzyme used to join DNA molecules together). The gene itself was from an *E. coli* virus. Genetic engineering had put the gene for the DNA ligase next to a very efficient promoter, so that this *E. coli* made huge amounts of DNA ligase. This project demonstrates how the results from a piece of recombinant DNA research can often produce new methods for doing further recombinant DNA research. In this case the project turned DNA ligase from an expensive, rare enzyme into one which was cheap and readily available. Several other enzymes for recombinant DNA technology are now produced in large amounts by similar methods. Thus the technology contributes to its own ever-increasing rate of growth.

At the other end of the size scale, ICI have taken the idea of using a bacterium to make a protein to one logical conclusion: they have developed a bacterium which simply produces a large amount of protein and which grows fast and efficiently on methanol, a cheap industrial solvent. These bacteria are then harvested, treated to make them chemically innocuous, and packaged as Pruteen, an animal feed. Their plant to make Pruteen went into operation in England in 1980. Few bacteria are suitable for such a process as they require carefully controlled growing conditions, or lots of expensive heating or cooling, or will only 'eat' expensive chemicals. ICI employed many genetic tricks in their search for a bacterium which had the best properties as part of a long-term project to use bacteria in the chemical processes with which they were already familiar. Without ICI's years of experience in running large chemical facilities this would have been a hopeless task; as it was, their plant – which was of quite modest size, making up to 1,000,000 tonnes of product a year – was a complex affair costing a total of about £50 million. The central vessel alone in which the bacteria were grown cost £7.5 million, and needed a computer to control all the valves and pipes leading into it and taking the bacteria into the 'downstream' processing where they were turned into food for cows.

ICI think that the plant is only marginally economic to run, partly because of the huge research investment (another £40 million over fourteen years), partly because of the rise in oil prices during the project which made both energy and methanol more expensive, and partly because of political problems. Pruteen was a direct competitor to US soya-bean-based feeds, and in the late 1970s, when ICI were campaigning for permission to build larger plants elsewhere in the world, the last thing that President Jimmy Carter wanted to encourage in the run-up to the

1980 presidential election was a competitor to one of his home state's major crops.

But while Pruteen is only a marginal product now, this could change if the oil price remains as low as it was during early 1986, or if the cost of farming in the United States rises substantially (which it might: we shall discuss this in Chapter 13). The same is true of many other potential products. Using genetically engineered bacteria to produce amino acids such as lysine is no more profitable now than are other methods, but a slight shift in the international chemicals market could alter that. The Japanese are the largest producers of amino acids like lysine, and they will not hesitate to grasp the new technology if it becomes profitable – their record in the micro-chip industry shows that. All the other products mentioned in this chapter are also made only in amounts that will satisfy specialist markets – there is no million-tonne international market for DNA ligase. But the technology of genetic engineering, and – equally important – the chemical engineering techniques to use the engineered bacteria, are nearing the maturity that would need only a small alteration in oil prices or a few minor developments in technique to make genetic engineering economic as a method for making large-scale products.

We seem to have stopped discussing the esoterica of gene splicing and instead to have spent the last few pages talking about money. This is inevitable. Many of the fundamental problems in making genes do what we want them to do have been overcome, and so with increasing frequency the question to ask is not 'Is it possible?', but 'Is it worth doing?' That is not to say that technical questions are now behind us: even in the relatively simple genetic engineering necessary to produce a single protein from an E. coli, there can be many unpleasant surprises for the researcher which can make a project more difficult or even impossible. But there is now a wealth of experience to draw upon, and so it is more likely that, if a project uses tried techniques, it will work. The next question is, of course, whether it will pay.

Along with the economic problems facing the future of industrial genetic engineering, a battery of legal ones has sprung up. These, too, are assuming an increasing importance in deciding which projects, applying genetic engineering to industrial processes, should be undertaken.

First came the problem of safety. In 1975 this was the concern uppermost in people's minds, as no one had much idea of what the new technology was able to do. All research was halted for a while in a moratorium on recombinant DNA, and subsequently all research had to be carried out in sealed laboratories where recombinant DNA was watched as closely as plutonium. Several cities, most prominently Cambridge, Massachusetts, which held Harvard University within its limits, debated whether they would allow the new research at all. A battery of strict regulations governing how researchers could carry out their experiments was set up, to the irritation of scientists who were in the process of proving recombinant DNA techniques to be extraordinarily safe. Lagging some years behind, legislatures eventually realized this too and relaxed most of the regulations. (We mentioned some of the evidence that contributed to this debate at the end of the last chapter.) Still, government regulation of genetic engineering has not vanished, and remains particularly strong when someone wants to take a genetically engineered bacterium outside the clean and isolated environment of the laboratory. This sometimes produces some peculiar results.

In 1984 two groups in the United States were planning to use recombinant DNA techniques to produce a bacterium (called *Pseudomonas syringae*) which lacked a particular protein. This protein was the 'nucleus' around which ice particles form in frost on some plants. Just as seeding a cloud with crystals of the right shape can encourage the formation of ice crystals to start a rain shower, so this protein encourages ice crystals to form on the surface of leaves if the air temperature falls far enough. As a result the bacteria encourage frost to form, damaging the crops. The aim was to replace the natural population of *Pseudomonas syringae*, which grow on some crop leaves, with engineered ones which could not produce this protein. Thus in cold weather there would be a reduced risk of frost damage to the leaves of the treated plants – the bacteria would be a form of frost-proofing. Both groups applied to the National Institute of Health (NIH) Recombinant DNA Advisory Committee (RAC) for approval to try some field trials, and both received it. But following a vigorous campaign by veteran anti-DNA campaigner Jeremy Rifkin, a judge overruled the RAC and told the NIH that its experiments should be stopped, as the study of the impact the new bacteria might have on the environment had not been presented. (Actually, the problem was a procedural one about how the studies had been presented, not that they had not been done.) The NIH ordered the federally funded research group run by Dr Stephen Lindow to hold up their trial experiment. The other group were from Advanced Genetic Sciences Inc., a commercial research company. Because they were not funded by the government, the NIH could only advise them on what to do, not order them, so their plans went ahead. The difference was not in the experiment, but in who was paying for it.

Not to be put off, Mr Rifkin tried again in 1986 and succeeded in getting the local government of Monterey County, California, where Advanced Genetic Sciences' experiment was to take place, to ban it. Only in 1987 was the 'frost-free' bacterium tested in realistic field conditions.

The irony is that 'frost-free' bacteria occur in nature anyway. Lindow's group found that a small percentage of natural *Pseudomonas syringae* had a mutation which destroyed the ice-protein gene. (This was one reason why the RAC did not consider the proposed field trial to be dangerous.) The research groups could have simply collected these natural mutants and thus would have had far fewer problems with regulatory bodies, who are less concerned with bacteria per se than with recombinant DNA, and presumably less difficulty with Jeremy Rifkin. However, the researchers chose to engineer their own version of the 'frost-free' bacterium not for scientific reasons – the bacteria are effectively identical to their natural counterparts – but because a genetically engineered bacterium would be patentable while a 'wild' one would not.

When genetic engineering was halted because it was considered potentially dangerous, scientists abided by the restrictions because they could see a very important reason to do so. But the restrictions faced by genetic engineers today are of a very different nature, and are felt by some scientists to be obstructions created for purely political, not technical, reasons. Whether this is true or not, it has had two results. A technical problem can be overcome or side-stepped. When it was believed that genetic engineering was dangerous, technical means (which we surveyed at the end of the last chapter) were used to make its practice safer. But there is no technical answer to a political problem. The only alternatives open to the scientists are to wait for the lawyers to sort it all out (a very expensive and time-consuming option), to give up entirely, or to bend the rules. Monsanto, a leader in the application of genetic engineering to plants, is already debating whether to give up in this area because of the legal pitfalls involved when trying to test any product. At least one company has bent the rules. Biologics were very liberal in their interpretation of the regulations governing the use of recombinant DNA outside the laboratory when they tested their pseudorabies vaccine on pigs (see page 106). They had ample *technical* evidence that what they were doing was entirely safe, but they were foiled by other rules. In exasperation, they took shortcuts.

However, a number of companies have been eager to use genetic engineering in preference to other methods not only because of the technical advantages it can offer but also for other legal reasons, this time involving the patent laws. The problem of patenting genetically engineered strains has been a tough one both for researchers and for lawyers. After many years of debate in successively higher courts the US Supreme Court upheld a patent granted to Herbert Boyer and Stanley Cohen (of UCSF and Stanford University respectively) which covered many of the basic techniques of genetic engineering. This did not go down well with the scientific community, as Boyer and Cohen, although at the forefront of the developers of the new technology in the early 1970s, were by no means its only inventors. Some of the objections were removed when they said that the patent would be held by their universities and that royalties would go to the institutions and not to the researchers themselves (although a few asked whether Stanford, with an annual income of around $200,000,000 a year, really needed the cash). In addition, no royalties would be charged to academic researchers. Even so, it is very unclear how the numerous techniques which Boyer and Cohen did not invent can be separated from those they did, when these two methods are in reality integral parts of one process. Nor is it certain whether the whole edifice of legal argument would stand up to an attack by a good set of corporate lawyers in court. It is probably only a matter of time before one of the major users of recombinant DNA methods decides that they just are not going to pay any more, and then the whole business will be up for grabs again.

A set of patents in a related area of biotechnology is already under attack. The production of 'monoclonal antibodies', another novel biotechnological method unrelated to recombinant DNA, was pioneered by Cesar Milstein in Cambridge, England. Neither the Medical Research Council, which funded the work, nor the National Enterprise Board, which supposedly exists to exploit it, wanted to patent the process until they realized that some US researchers had already taken out a number of key patents. This generated enormous resentment among researchers who usually get by on the principles of fair play without a battery of legal safeguards, especially as it was almost universally agreed that it was Milstein's technique and, if anyone, he himself should patent it. Because of this, most laboratories ignore the patent, which has not been defended in court and so is likely to vanish as a curious but ineffective piece of legal history.

These are attempts to patent techniques so broad that entire research institutes do nothing but apply them, so perhaps it is not surprising that they are running into legal heavy weather. Surely it would be easier to patent a specific product, say a genetically engineered plasmid? Not so: the researchers trying to get patent protection for their genetically engineered bacteria came across just as many problems to start with. The genetically engineered *E. coli* was a living thing, and as such could not be patented (a specific and rather controversial exception had been made in the United States to cover plant varieties; but the law in Europe does not allow plants to be patented). However, DNA was not in itself a living thing, and the core of the technique was the alteration of DNA to manipulate the information in the genes.

Information in the form of a book *could* be copyrighted. In 1980 a court decision clarified the position on the ownership of computer programs, giving the program's developers rights over the program and any reasonably minor variation of it. In the wake of that it was decided that genetically engineered DNA was like a computer program and therefore could be patented. Hence the decision of Dr Lindow and Advanced Genetic Sciences to make the genetically engineered 'frost-free' bacterium mentioned above: the normal bacteria would be unpatentable, while the recombinant DNA used to make the genetically engineered version could be the subject of a patent. This decision was important because, as we have seen, genes are simply information and information is very easy to steal. In spy thrillers, it is the plans for the submarine or the nuclear reactor which are stolen, not the object itself. In the early days of the race to clone the interferon gene, the following story had wide though unofficial circulation. When Biogen announced at a press conference that they had cloned the gene, they showed a slide of the complete sequence of the bases in the DNA they had isolated. A scientist from a major pharmaceutical company was there at the press conference, photographed the slide, took the photograph back to the company laboratories and used chemical reactions to make a piece of DNA with the same sequence of bases as Biogen's own gene. Biogen were probably not worried about that eventuality as they published the base sequence in a major scientific journal soon afterwards anyway. But whether it is true or not, the incident illustrated both the pressure to achieve results and the ease with which such results may be stolen in the absence of proper legal protection.

But can you actually patent an interferon-making *E. coli*? After all, Biogen did not invent interferon, or the plasmid into which they spliced the gene, or the *E. coli* in which they placed it. 'All' they did was to rearrange pre-existing information in ways which any scientist of the time could have predicted would produce the result they sought. What if someone else did exactly the same thing from scratch, producing the same result but using none of Biogen's information (as indeed happened, although all the other groups were a bit behind Biogen)? Where would Biogen's rights to cloned interferon genes end and the public domain begin? With their particular clone? With all similar clones, no matter who made them? With all clones derived from the Biogen original? Most genetic engineering companies have very competent legal departments!

Another grey area is the boundary between industry and academe. When Genentech produced their own interferon-making *E. coli*, it had a varied history. In the early days of interferon research, the University of California at Los Angeles had produced a culture of human cells which made a lot of interferon – not a large amount by medical standards, but more than was usual for human cells. This meant that they possessed a lot of interferon RNA. The cell line was subsequently studied and improved by Hoffman-La Roche, and then by the independent Roche Institute. In 1980 Hoffman-La Roche funded Genentech to use the RNA from these cells to make cDNA and to clone that cDNA in *E. coli*. Now, who exactly owns rights to the resulting *E. coli*? Without the cells, the project would have been much harder. Without Genentech's gene-cloning expertise it would also have been very difficult. Without Hoffman-La Roche's expertise in bulk drug production, the Genentech bacterium would have been useless commercially. So should everyone in the story have their names on the product? By the end of 1985 the US courts had come to only one conclusion: the relatives of the patient who had originally supplied the cells (and who subsequently died of cancer) had no right to them or to the interferon they would eventually lead to.

These problems are not unique to genetic engineering. The ownership of strains of bacteria and yeast has been a tricky problem for the brewing and baking industries for centuries, and chemists and nuclear physicists have had to reconcile having one foot in industry and one in academe long before biologists were faced with this dichotomy. But the unique portability of molecular biological ideas, the ease with which a gene sequence may be stolen, and of course the fact that once you have a billionth of a gram of a plasmid you can put it into an *E. coli* and grow as much as you want, meant that ownership in genetic engineering is a tricky thing to enforce.

In this chapter we have seen the first applications of genetic engineering. Only a handful have made it to marketable products, although dozens of hopeful projects are in the pipeline. The successful ones had a few features in common. They aimed at cloning the gene for a single protein, one whose identity and function were already known and which could work well when isolated from other proteins. The aim was always to make a bacterium that would make more protein. This type of project will continue for many years yet. We have barely scratched the surface of the treasure trove of regulatory factors in our bodies which may have medical use; as genetically engineered bacteria are produced which make these proteins, we will gain more insight into which ones are potentially useful to medicine. The number of processes which could usefully be speeded up by an enzyme is vast. As genetic engineering techniques become more adept, it will become increasingly likely that these projects will be economically viable. At the moment only the biomedical projects are reasonably sure to show a profit (if they can overcome the numerous technical hurdles and still demonstrate that their product does some of what it is meant to do). As research finds cheaper ways to grow bacteria in bulk, and if the costs of more traditional industry rise, we should see less glamorous projects generating fewer spectacularly expensive products. Although less popular with the media, such projects will affect the lives of far more people.

But there is one other success story from the early days of genetic engineering which we have not yet touched on, because it is not a commercial project at all. Is it possible for a project to be quite successful, to hold great promise for extensive future applications and yet to hold out little hope of profits for any research company? It is. The use of genetic engineering to diagnose inherited disease is such an application, and one we shall discuss in the next chapter.

Chapter 9

The Genetic Detectives

In Chapter 5 we mentioned a number of diseases caused by defective genes, such as thalassemia, which is caused by a defect in haemoglobin genes, and several others, like diabetes, in which genes are believed to play a role. One of the hopes of the new technology of recombinant DNA was to have a more direct way of diagnosing and treating such diseases. The treatment – 'gene therapy' – we shall leave to the next two chapters. However, the diagnostic aspects of the new technology have borne a greater wealth of fruit than even their most enthusiastic supporters would have predicted ten years ago. Characteristically, therefore, those supporters are now predicting more from the technology than it can possibly deliver.

The diagnostic side of the methods are all based on a single technique called 'nucleic acid hybridization', or 'hybridization' for short. Like the self-replication of DNA and the way that sticky ends bond together, this technique makes use of the way that the bases in the DNA chain tend to hold together in specific pairs – A always with T, G always with C (see page 18). Hybridization is just another facet of this propensity of DNAs.

If we separate the two chains of a double helix, they will tend to come together again. They do this because the bases in the two chains are complementary – indeed, the fact that they are complementary and hence tend to stick together holds the double helix together in the first place. But what if we separate them and prevent them from joining up again? They will still be in a more stable state if they are stuck to complementary bases, and so will 'hunt' for a set of complementary bases. If there are no others around, of course, then our DNA will remain as single, isolated chains. But in the hybridization technique we arrange for there to be a large number of other DNA chains around. One of them is complementary to our original DNA chain, although we do not know which. All the others are not complementary. Eventually our original DNA chain will find that one complementary partner and join up with it to form a double helix. The double helix is so stable a structure that it can form even if the two strands are not perfectly complementary: exactly how many mismatches will be allowed depends on the conditions under which we perform the experiment, but is typically no more than 10 per cent. The result is a hybrid molecule, a double helix in which one of the DNA strands comes from our original DNA molecule and the other comes from somewhere else entirely. Hence the name 'hybridization'.

This is a very useful way of using one piece of DNA to 'find' another. The DNA with which we start can be made radioactive – this piece is called the 'probe'. Then we separate the two chains of a lot of copies of our probe DNA and add them to a whole lot of other DNAs which we have separated physically but have not identified. Some of the probe DNA will hybridize with any of these DNAs which have a complementary sequence – that is, it will form a double helix with them. It will not touch any of the other DNAs, so when we remove all the remaining probe which has not formed into double helices, the probe which is left behind will be that which has formed a hybrid double helix. As the probe is radioactive, this double helix will also be radioactive, and we can readily detect in which of our DNA samples it is. Thus at the start of the procedure we had a lot of DNA samples all of which were chemically similar, but at the end we have one which is radioactive while the rest are not. That single radioactive sample is the one which contains DNA whose base sequence is complementary to the sequence of the probe.

To make this work, the probe must be complementary to only a few DNA sequences, as otherwise bits of DNA would be 'labelled' by our probe in every sample. And, of course, we need a piece of DNA which is complementary to at least one of our DNA samples, or else we will end up with all our samples just the same as they were when we started!

This technique is not in itself part of the battery of methods which create and manipulate recombinant DNA. However, it is closely linked to those techniques, because nucleic acid hybridization works best if we have a pure DNA for use as a probe. Just any old DNA is no use; if the probe is to tell us anything useful about our samples it must be pure, so that we can say that sample number 76 contains a base sequence complementary to, for instance, the Lac genes, and that none of the other samples contains that base sequence. To say that sample 76 contains DNA whose bases are complementary to either the Lac genes or one of three dozen other families of genes is not so helpful! One of the results from work on recombinant DNA is the production of pure DNAs – plasmids containing only one gene. These can then be used as probes for a variety of hybridization experiments. Indeed, so closely are the techniques of gene cloning and hybridization interlinked that together they form the mainstay of research into how genes are put together and function.

So effective is hybridization as a method for detecting specific DNA molecules that it can be used to detect one molecule which is complementary to a probe from among a million non-complementary ones. For example, it can be used to find a single gene clone among a huge collection of them, as we mentioned on page 82. It can also be used more directly to detect genes in the DNA of our cells. Once a gene has been cloned, we can use that isolated gene clone to prepare a probe. This probe can then be used to detect the presence of any similar gene in other DNA, for example the DNA from a suspected carrier of a genetic disease. There is no need to clone the DNA from the suspected carrier to do this: we can make DNA from the cells in a few millilitres of blood and perform a hybridization test on it to see if the mutant gene is present. Even more cunning is a combination of this method with the use of restriction enzymes. Some mutations are caused not by the lack of an entire gene, but only by the lack of a gene which makes sense. A single base in the DNA chain may have got out of place, rendering the message of the gene useless. If that base change occurs at the site at which a restriction enzyme cuts the DNA (and there are a lot of known restriction enzymes which cut the DNA at a large number of sites), then the restriction enzyme will no longer recognize that particular site and so will fail to cut the DNA there. Thus the size of the pieces into which a specific gene will be cut by that particular restriction enzyme will be altered. Using our probe, we can detect just which pieces (out of the million or so that this restriction enzyme will generate from a whole cell-full of DNA) are the ones that come from our gene. Variations in these methods enable us to discover whether parts of genes have been swapped around or removed, or whether the whole gene has been removed.

The neatest method for using hybridization, though, is to look for one specific mutation. If we know approximately where the mutation is, we can manufacture a probe by making a short piece of DNA or RNA and, by carrying out our hybridization under specific conditions, detect whether that specific mutation is there or not, regardless of whether it affects a restriction enzyme's cutting pattern. This method has been applied on a laboratory scale to the detection of carriers of the mutation which causes sickle cell disease, another disorder of the blood.

These methods are dependent on recombinant DNA techniques and, as we said earlier, those techniques are not limited to the study of particular genes: they work on DNA, and all our genes are stored on DNA. So in principle we could use these methods to detect any of the mutant genes we mentioned in Chapter 5. In practice, however, there is a hitch. The hitch is that we need a probe. If we know what the base sequence of our target gene is (which usually requires us to have a gene clone of that gene), then we can make our probe knowing what its base sequence ought to be, or we can use the cloned DNA directly. But either way we need to have our gene cloned first.

That raises a problem. If we know what gene is defective in a particular disease, then 'all' we have to do is clone that gene and we can make a probe. Such is the case with thalassemia because, as we mentioned in Chapter 5, we know exactly which gene is defective. But in many other cases no one knows what causes the disease. For example, we know that Duchenne muscular dystrophy is caused by mutations in a single sex-linked gene. But no one knows what that gene does in normal people. Does it produce an enzyme, a structural protein, a regulatory factor? No one can tell, but without that knowledge scientists cannot isolate a clone of that gene directly, as we are faced with a million gene clones only one of which is the one we want. It is a paradox: until we clone the gene, we will not be able to find out what it does, and until we have some idea of what it does, we will not be able to clone it!

Well, there is a way round the paradox. We can clone a piece of DNA from a nearby gene, and then use that gene clone as a probe to clone another, even closer, piece of DNA and then use that as a probe to get closer still . . . and so on until we arrive at the gene we want. This technique is known as 'chromosome walking', as we are edging along the DNA, the chromosome, in small steps to arrive at where we want to be. It is a long and expensive process, but at the moment it is the only way to get at those vital genes.

This is not quite as unrewarding as it may sound, because those nearby pieces of DNA can sometimes be of use. They can be used as probes in their own right – not for the gene we want to detect, but for pieces of DNA very near to that gene on the chromosome. We can be fairly sure that if a suspected carrier inherits such a piece of DNA from one parent, then he or she will also inherit the gene next to it in the same parent. So we can follow around the piece of DNA *next to* the gene in lieu of being able to follow the gene itself. This is not foolproof, and the further away our 'nearby' DNA is the less foolproof it becomes. Nevertheless it is a start. Such nearby pieces of DNA are called 'linked markers', as they mark a place on the chromosome which is physically linked to the gene we are interested in studying.

This technique of chromosome walking has been the method of choice for groups trying to obtain cloned probes for genes whose defect can cause disease, and as a result several of them have now generated linked markers to diagnose these defective genes in potential disease patients or carriers. Among the genes for which such diagnostic methods are available now are muscular dystrophy, cystic fibrosis and Huntington's chorea. It is hoped that possession of the linked marker DNAs will lead to cloning the actual genes involved in these diseases within the next year or so. Indeed, in October 1986 such an exercise in chromosome walking appeared to have reached the gene for muscular dystrophy. However, this still needs to be checked.

But for two problems, linked markers could be used today as the basis of a system for diagnosing carriers of genetic diseases. The first problem is that these techniques are all very expensive to perform, requiring highly trained people both to perform the tests and analyse the results, and there are not enough such people to implement genetic tests, using recombinant DNA, on a national scale. So investment of money and manpower is needed on quite a large scale to make the tests widely available. That would be a fairly simple political decision if it were not for another dilemma: the ethical problem. The difficulty is that the tests are far from infallible. No medical diagnosis is infallible, of course, but linked markers are subject to rather more uncertainty than many. Should this uncertainty be foisted on the patients? A recent case highlighted the problem. A group in the United States has cloned a linked marker which is very close to the gene for Huntington's chorea. Although this cloned piece of DNA is very close to the gene whose defect causes the disease, it is by no means just next door to it; consequently using it for diagnosis is likely to lead to a number of cases where someone is identified as carrying the defective gene when in fact they are not – and of the converse, a failure to diagnose a true case. Is it ethically right to use this linked marker DNA as a probe to try to diagnose Huntington's chorea in teenagers, knowing that some of the people who will be told they have the disease, and whose lives will be devastated as a result (occasionally people commit suicide on hearing similar verdicts of doom on their medical future), will in reality be perfectly normal? James Gusella of the Massachusetts General Hospital, the head of the group which cloned a probe which hybridized to this particular linked marker, thinks that such a use is not ethical, and will not give the probe to researchers who want to use the gene clone for diagnosis. An Oxford group thinks that the people on whom it wants to use this probe are intelligent enough to understand the uncertainties involved, and that they should be able to employ the probe for medical use. With hard work and a bit of luck, this particular ethical dilemma will be side-stepped in a few years by the identification of better probes. But it is not unique to this disease: genetic diagnosis using cloned genes is seldom straightforward, and will often be subject to a degree of uncertainty. Whether patients should be exposed to this uncertainty is still something on which the medical profession has not reached a conclusion.

This technology is receiving wide coverage in the press nowadays because there are no technical reasons why it should not be applied to any disease in which genes play a major role. Using these probes, we could detect all the carriers of the recessive mutations we mentioned in Chapter 5, thus removing untold misery among familes afflicted not merely with the diseases themselves, horrible though they are, but also with the uncertainty about where they are going to strike next. Because of the potential for good, a tremendous effort is being expended to clone the genes whose defects cause cystic fibrosis, muscular dystrophy (as I mentioned, at the time of writing this one has probably been cracked), Huntington's chorea and others. Many other genes for diseases which are quite rare in the Western world have already been cloned. A major problem in these cases is how to scale up these expensive and technically tricky tests so that they can be used on all the people who want them, and here at last the biotechnology companies are beginning to take an interest. Simple kits which can be used by any hospital for genetic screening have the potential for major profits, something which the diagnostic side of genetics has lacked so far. It is on the production of such kits that biotechnical companies are concentrating. When these techniques are perfected, the developed countries will have the potential to eradicate such diseases almost entirely (muscular dystrophy may be an exception to this). Of course, no matter how streamlined the test, eradication will be expensive – tens of millions of pounds for *each disease* in the United Kingdom alone. Whether this is politically acceptable or not is another matter. Again, we are back to talking money and politics rather than genetic engineering.

Who can argue with a teenager who wants to know whether their life is going to be blighted by Huntington's chorea, or a mother who wants to know the risk that her child will suffer from cystic fibrosis? But this is only the most straightforward application of the technology. Here, in its most dramatic form, genetic engineering overcomes technical barriers to face ethical issues which, although of great importance in deciding what *should* be done, are not the point in a discussion of what technically can be done. So let us turn for a moment from those few diseases whose genetic diagnosis is within the grasp of present technology, and see to what other aspects of our health these techniques could be applied.

We mentioned that everything we are and do is affected to some extent by our genes. Many diseases are affected very little by our genes – nearly everyone catches colds, for example, and the major cause of colds are the varieties of cold virus to which we are exposed and the general level of our resistance when we encounter them, not our genes. But other diseases are much more prone to genetic influence. Heart disease and cancer both run in families – this is why life insurance companies want to know whether our parents died from these diseases before they work out our life expectancy, and hence our premium rates. This can be translated in genetic terms to there being one, or probably several, genes that dispose us to these diseases. Depression may be another disease which is strongly influenced by genes (a research group may now have found one of the genes which can predispose some of us to clinical depression), and diabetes is almost certain to be caused in part by our genetic inheritance (although the gene for insulin seems not to be involved, even though diabetes is a disease caused by the body's failure to make insulin). Now at present no one can prove that these 'predisposing genes' really exist, as other, environmental, causes could be envisaged for diseases running in families. If a family inherits money, it may eat a richer diet and so be more prone to heart attacks: the cause of the family's high heart-attack rate is therefore a combination of plenty of money and what it is spent on, and not even the most ardent supporter of the 'nature' school of the 'nature vs. nurture' debate would argue that money is part of our genes. However, many of the studies which uncovered such potential genetic effects tried to balance out the effects of different environments, so it seems likely that many diseases truly are affected by our genes.

At present no one knows what those predisposing genes might be doing. But then no one knows what the gene for muscular dystrophy does either, but that does not stop scientists from trying to track it down, using the technique of chromosome walking; we could imagine that the same methods could be applied to tracking down the genes predisposing us to diabetes. And then the whole population becomes the target for screening to see if they are at risk from diabetes. We should not underestimate the problems in doing this: this is not a project for the next year – more likely for the start of the next century. There are almost certainly several genes involved, none of which has been identified, and whose effects are mixed up with the effects of the environment. Even when the effect of one gene has been successfully identified, scientists will have to clone that gene by the laborious process of chromosome walking and then will have to prove that the gene they end up with actually does what they believe it to do. Only then will it be of any use in finding out why some people develop diabetes and some do not, and in predicting which are at risk.

So while the methods are known, the actual research project required would be enormous. And it may turn out that there are so many genes and so many environmental influences involved that the whole project becomes unworkably complicated. We are at too early a stage to tell. In 1973 all anyone engaged on interferon research could say was that it was possible in principle to get a bacterium to make interferon, but whether it was possible in practice or whether the result would be of any use was anyone's guess. We are at the same stage in much of the genetic diagnosis field: if the genes are there, it is possible in principle to use them as probes. But whether they are there, and whether the probes would have any practical application, is unknown.

If the genes are there, then again the technological questions will start to take second place to social ones. Would you want to know if you were prone to diabetes so that, if you were, you could keep to a moderate diet and avoid developing the disease completely? I would; some would not. Would you want someone to screen you and tell you that you are prone to schizophrenia and must undergo preventative psychiatric therapy? I would not; maybe you would. The same technology could lead to the halving of the incidence of heart disease, cancer and premature senility in our country, or to supermarket shelves urging pregnant women to 'Give Yourself the Baby You *Deserve*', with do-it-yourself abortion kits if the growing fetus's genes predict that it will have brown eyes instead of blue. At the moment such a supermarket kit would cost tens of thousands of pounds and for its use would need a licence for handling radioactive materials (remember our probe was radioactive), so it is not a realistic option. But alternatives to the radioactivity are already being developed – a few more technical advances could make it plausible, maybe even economic.

But my biases are taking over what started out as a discussion of technology. Let us return to technology.

So far we have talked about the use of genetic engineering of bacteria and yeasts, both to produce products which we can use in industry and to produce large amounts of DNA which we can use in medical diagnosis. But there is another option we have left unexplored: the genetic engineering of higher plants and animals. While this chapter has addressed human welfare, as indeed had the last, we must now move on to those aspects of genetic engineering which aim to alter the genes of organisms more complex than bacteria, including our own.

Chapter 10
The Problem with Plants (and Animals)

Traditionally, the phrase 'genetic engineering' has been applied to the genetic manipulation of higher organisms. Doctors Frankenstein and Moreau did not mess about with *E. coli*; they started at the top with artificial men. Much of the concern over the legal and moral implications of recombinant DNA techniques is not focused on *E. coli* but on the possibility that the technologies now being applied to *E. coli* will one day be applied to us. So what are the chances?

There is a simple answer : eventually someone will be able to apply a form of genetic engineering to human beings. That was never really in doubt since biology developed from a descriptive science into a manipulative one. But what are the chances that today's technology will be used to manipulate higher animals and plants in the foreseeable future? That is a more complex question. In this chapter we shall look at the problems and the methods in the field, and in the next we shall examine what has been achieved.

DNA is the (almost) ubiquitous information-carrier of life. Some of that information is concerned with directing the cell to produce a protein, using the genetic code, while other regions of DNA control how the DNA itself is used. These are the generalities of what we have surveyed in the last few chapters, but within that framework there is plenty of room for variation. There are major differences between the organization of the DNA in bacteria and in higher organisms, and as these differences are crucial to the success or failure of genetic engineering of animals and plants we should examine them briefly.

At the most obvious level, our cells have a nucleus while bacterial cells do not. For a broad understanding of how genes work this has not mattered very much, but when we try to alter the genes in those cells the differences become more important. The presence or absence of a nucleus is just the most obvious manifestation of quite major differences in organization between the DNA of *E. coli* and ourselves. In *E. coli* the DNA lies as a coiled heap, like a large bale of wire, near the centre of the cell. Various proteins hold it in place, and others, enzymes like RNA polymerase (page 67), are latching on and off so that they can read the information in the sequence of bases. However, substantial amounts of the DNA are uncluttered by proteins.

The situation is rather different in the nuclei of our cells, and those of plants and fungi too. Here, the DNA is packed with proteins called 'histones' which bend the double helix round on itself to form a new, larger coil. This coiled DNA is then folded into larger coils, and those into still larger ones. The whole nucleus is simply the highest level of this complex packaging system – the outer case for all the coils. The nucleus of a human cell is therefore physically structured like a warehouse, in which toys are packed into boxes, themselves packed into crates which are piled up on pallets which are racked on shelves inside the building. But unlike a warehouse, where everything is kept in discrete packages, the DNA in the human nucleus runs as a vast thread through the whole assembly, so in the resulting complex of DNA and its protein packing it is very difficult to work out where a particular piece of DNA is heading next. The cell has it all under control, of course; indeed, one of the reasons put forward to explain why the DNA in animal cells is in such a convoluted state is that there is just too much of it to store in any other way, if the cell is to have instant access to any gene. Like a road map, it needs to be folded back on itself again and again simply to keep the whole thing manageable.

This tight packaging of DNA into our nuclei is not only for storage. Scientists have found that the packing of sections of DNA is subtly different if the gene encoded by that section of DNA is actually being used when they examine it. The structure of the packing of genes which are being used is a little more open than that around genes which are not being used in a particular cell, and slightly different proteins are used to pack the DNA into its tightly coiled conformation. This is quite understandable, as it is indeed hard to imagine how the cell could read the information encoded in DNA if it were tightly packed around itself like boxes in a full warehouse. However, it does raise the possibility of a problem for the genetic engineer. Not only has he to get the right gene into the cell, but he will have to ensure that it is packed in the right proteins if it is ever to be used in that cell.

The organization of DNA is somewhat different in bacterial and mammalian cells, too: their 'filing systems' for the genetic databank are not really compatible. In E. coli nearly all the genes are on just one long segment of DNA, and it is circular: it has no free ends. In man there are forty-six long segments, each containing far more DNA than E. coli's single loop, and they are straight – their ends are free. In E. coli the large loop may be accompanied by smaller pieces called plasmids, but in higher organisms there are few examples of such separate pieces of DNA in normal cells. Yeast cells have a few, some plant cells may have some and a few animal viruses actually get their DNA into their target cells as a closed loop which can behave rather like a plasmid once it is there. The Papilloma viruses, which cause warts, are of this sort – the possibility that having such independent circles of genes in mammalian cells could be a bad thing is illustrated by the fact that a type of Papilloma virus, as well as causing warts, has been implicated as a cause of cervical cancer in women.

In the cells of higher organisms these smaller circles of DNA are also found packaged into compact coils with the histone proteins. In fact, if we were to put any DNA into a mammal's cell which was not packaged in this way, it would either be packed up quickly by the cell or else would be broken down into its component bases. The cell does not like its DNA lying around with no proteins on.

Another aspect of the difference between E. coli and man is the amount of DNA in their respective cells, and what it is used for. Our cells contain 6,600,000,000 bases of DNA each, compared with 4,200,000 in E. coli. As we mentioned in Chapter 6, one of the reasons that we have more DNA than E. coli is that we need more genes. Scientists expected humans to have more genes than a comparatively simple bacterium. But when the focus of genetics shifted from studying the effects of genes in breeding experiments to studying DNA itself, a problem arose which has not yet been sorted out. We have about 750 times as much DNA as *E. coli*, but we do not seem to have 750 times as many genes. Even odder are the amounts of DNA sported by some species. Some types of frogs and newts have ten times as much DNA as we do, although it is hard to see why they would want ten times as many genes: the record is held by the lungfish, which has about fifty times more DNA in each of its cells than we do, although zoologists consider it a rather lowly and primitive organism. So it appears that the amount of DNA in a cell is not a good guide to the number of genes which that organism is likely to require. This in turn suggests that not all the DNA in a cell need be concerned with information storage.

So what is the 'extra' DNA for?

We still do not know. The function of most of our DNA is a mystery. Indeed, much of it seems to be so useless that one theory has been developed which says that it really *is* completely useless – it is, in the terminology popularized by Leslie Orgel and Francis Crick, 'junk DNA', DNA with as much use as the reverse side of a picture. However, despite the backing of such an influential theorist as Francis Crick, the 'junk DNA' school of thought has not gained universal acceptance. One reason for its rejection in some quarters is that a function is found every now and again for a piece of DNA previously thought to be completely useless. But also it offends some biologists' sense of 'rightness' that more than half our DNA should be completely useless! The existence of this 'extra' DNA makes the job of a genetic engineer rather harder, because that DNA may have a crucial role which we have yet to discover; we may only find out what it is when we try to engineer those genes and something goes wrong. We could be missing a vital piece of information on what effect our manipulations might have. We would in fact be in the position of cavemen trying to repair an abandoned automobile: they would have no idea what the gas tank is for – it is, after all, a completely empty space, apparently useless – but the automobile will never go if we ignore the gas tank.

In a few cases, functions are being assigned to some of the 'extra' DNA, and quite a lot more of it is being accurately charted and described. One curious finding of such recombinant DNA research is that animal and plant genes are organized differently from bacterial genes. In bacteria a gene will have a promoter on its front end, followed by a length of DNA which instructs the cell how to make a protein. The first three bases indicate what the first amino acid should be, the next three what the second amino acid should be, and so on. But in higher organisms most of the genes are not continuous stretches of useful information. Instead they have other pieces of DNA which carry completely different information – or (maybe) no information at all – inserted into them, like commercial breaks in a television program. These 'commercials' can get so out of hand that in some genes over 90 per cent of their total length is 'commercial' and only 10 per cent actually carries information for making a protein. In some cases the gene can still function if the inserted DNA is completely removed, so maybe this DNA really *is* useless. But in others it seems to be necessary: a number of the genes for antibodies have gaps of this sort containing a segment of DNA which is responsible for controlling how fast that antibody gene works; in another case, this time in yeast cells, a gap in one gene contains another, entirely different, gene. It is as if a novel by Dickens had the whole of *Macbeth* inserted between Chapters 8 and 9.

In order to make sense of these genes, the cell must get rid of the gaps. It does this by 'editing them out' when it makes the RNA, so that the RNA can be 'read' to make a protein. Thus the 'commercial breaks' are only a feature of the DNA, not of the RNA. This is one reason why it is better to use a cDNA – a DNA copied from an RNA – to produce a gene clone which is to instruct E. *coli* to make a protein. If that gene clone had 'commercial breaks' in it, it would be useless for this purpose, while a cDNA, with no gaps, would do the trick.

So there is a whole lot of our DNA which may be vitally important to controlling how our genes work, but if it is, no one has found out in what way. But this is not all a potential genetic engineer has to face. There are still all the controller genes which *have* been identified by recombinant DNA experiments. Genes in mammals usually start with a promoter. They can also have a segment of DNA called an 'enhancer' which, reasonably enough, enhances the activity of a gene – that is, makes it work faster (the segment of DNA in the middle of the antibody gene mentioned in the last paragraph is of this sort). Enhancers, once thought rather exotic features of virus DNA, are now being found in all sorts of other genes, and may turn out to be very important to the control of genes in animals. Then there are other control segments which respond to hormones, and others again which make a gene active in a particular cell, such as those in muscle, while it is not active in others, such as those in bone, and so on. Other regions are responsible for targeting the histone proteins so that the whole gene is packaged up in the right way. The list of bits of DNA which are important to the functioning of a gene seems to get longer every year.

If you find all this a little confusing, you are in good company. Molecular biologists are at a loss to explain how it all fits together, especially as they probably have only a tiny fraction of the whole puzzle to hand. The sheer size of genes in animals makes studying them difficult, and in the few cases where a gene has been really thoroughly studied so that scientists have some idea of how it works, it always turns out to be enormously complicated.

This, however, is only the problem of how the genes already in our cells work. A further problem faces a genetic engineer who wishes to place a 'new' gene in our cells. If we took a gene that we had cloned using recombinant DNA techniques and put it into a cell in a growing embryo (hardly an easy thing to do in itself), then it could have one of two fates. It could either join in with the DNA already in the cell, or remain aloof from it as a separate segment. DNA which joins in with the rest of the DNA in a cell must do so by linking into the middle of a chromosome: the DNA in our nuclei is arranged into chromosomes, and DNA which joins on to the end of a chromosome is subject to very rapid mutation. Thus we would want this DNA to be joined in the centre of the other DNA, just as we used DNA ligase to connect a piece of DNA with a vector plasmid (see Chapter 7). Once this has been achieved the DNA is said to be 'integrated', because it is intimately united with the rest of the DNA around it. We cannot use DNA ligase on our cells in the same way as we can on a DNA molecule in a test-tube; we would have to insert a DNA-cutting enzyme, our cloned DNA and DNA ligase into the nucleus of a cell (which would be very difficult), and then make sure that the enzymes only cut and joined exactly the right pieces of DNA and no others (which would be impossible, as all DNA would look alike to these enzymes). However, sometimes a cell will do this manoeuvre on its own, so it is possible to integrate DNA in this fashion.

Once a DNA is integrated into a cell's chromosomes, we might ask whether it is working as it should. Such experiments have actually been performed, and whether the 'new' gene works properly or not seems to depend in part on where in a chromosome it ends up. (We shall mention these experiments again in the next chapter, as they are examples of 'gene therapy' – the use of cloned DNA to alter the genetic function of animals.) The fact that the exact placing of our DNA in the cell's own chromosome can affect how the 'new' gene works would not be unexpected if you view DNA as a filing cabinet full of information. In any large filing system in which all the pages are of the same shape and size (true of DNA because the *bases* are always the same, it being the order of bases which is important), the meaning of a page will depend heavily on its context. Thus a page entitled 'Ignition Timing' will have a specific meaning if found in a file called 'Maintenance of your Ford', but a completely different implication if found in 'Arson for Beginners'. But such analogies are often misleading when used to predict what we will find in the real world, and no one actually expected DNA to be quite so like a filing cabinet in this regard.

The other option is for the DNA to remain separate from the rest of the DNA in the cell's nucleus. In this case, of course, our 'new' gene has no surrounding DNA to affect how it works. This requires some rather different genetic engineering, and specifically means that we must splice our 'new' gene into a vector molecule, as we did when putting it into an *E. coli* cell in the first place. We need a different vector, of course, as plasmids which are duplicated effectively in *E.coli* are not usually duplicated by the enzymes in human cells, and vice versa. However, once such splicing has been done and the genes have been hooked up to a suitable promoter (and often an enhancer, too, to boost their activity), then they will work, whatever cells we put them into. This is fine if we just want to produce a protein, but we can already do that with *E. coli*; the whole point of engineering more complicated organisms is to perform more subtle alterations. Also, the vector molecules can have the side-effect of giving the cells in which they reside some of the characteristics of cancer cells (recall that some *Papilloma* viruses, a type of virus whose DNA behaves like a plasmid, are associated with the cause of cervical cancer in women). So this approach is adequate for using human cells to make interferon, but we can already get *E. coli* to do that for us. It is not of so much use for genetically engineering whole organisms.

Perhaps it is time to turn away from the complexities of what we do not know and look instead at the relative simplicity of how we do not know it. The story of how we have come to understand something of the workings of genes in higher animals is a fascinating one, although rather removed from the theme of this book. The outline is important, however, for these results are almost all the products of genetic engineering experiments, and in turn limit what we can now do with recombinant DNA techniques. The knowledge rests on recombinant DNA technology's ability to clone a gene, to splice one short segment of DNA into a vector molecule in *E. coli* and then grow it independently of all the other DNA in the cell. In this way scientists can prepare large amounts of the DNA carrying the gene for, say, insulin, and isolate that DNA from all the other DNA normally present in a human cell (in Chapter 7 we outlined how this sort of genetic engineering was done); the result is that we can obtain the insulin gene, which is less than one millionth of all the DNA in a human nucleus, free from the other 999,999 millionths.

Methods used to study this pure DNA molecule, this 'pure gene', include using hybridization (see Chapter 9) to find out which RNAs might be made from various parts of the DNA, and to discover whether any of the DNA is similar to other genes in the same or different organisms. Scientists can also determine the sequence of bases in the DNA molecule and put that molecule back into living cells, with or without specific alterations to see what effects those alterations might have on the functioning of the gene. This latter process is very like causing a mutation to occur – we are altering the order of bases, the information content, of the DNA. However, by doing so on the pure DNA in a test-tube we can make sure that we know exactly what changes we are introducing, without the risk that changes will occur in other genes by accident. In this way we can build up a picture of the function of each of the regions of a gene. By analogy, we might conduct the experiment of taking the last sentence, chopping bits out and then giving them to a reader to see if they still make sense. We would discover that 'In this way' is dispensable for the broad meaning of the sentence, but 'picture' is not.

One of the most amazing advances in molecular biology has been in the ability to determine the sequence of the bases in a bit of DNA ('to sequence' has become a verb in its own right because it is such a common activity, like 'to clone' before it). Twenty years ago it took a whole laboratory months just to find out the order of the half dozen bases at the end of a DNA molecule; now a single molecular biologist can 'sequence' DNA at the rate of several hundred bases a day. The staff of the Cambridge (England) laboratory of Frederick Sanger, who shared the Nobel Prize in 1982 with Walter Gilbert and Alan Maxam for their development of this technology, has recently finished determining the base sequence of the DNA of Epstein Barr virus, a single circle of DNA 172,000 bases long. It took them about four years. They are now starting on another virus, cytomegalovirus (which we mentioned on page 48 as the cause of a rare type of cancer), whose DNA is over 300,000 bases long, and they expect to have it finished rather more quickly. 172,000 bases would fill about eighty sides of a paperback with pages packed with A, G, C and T. The US Department of Energy is starting serious discussions about the possibility of sequencing *all* the DNA in a human cell – 6,600,000,000 bases! They estimate it will cost between 300 and 1,000 million dollars. (For comparison, landing the first man on the moon cost the equivalent of 90,000 million dollars at 1986 values.) It is feasible, but is it worth it?

All this led to an information explosion which has left molecular biologists floundering. How on earth do you analyse 172,000 bases of DNA? Obviously not by going through it one base at a time 'by eye': you need a computer.

We started this book by comparing the computer revolution with the biological revolution, a comparison in which biology came off rather the worse. But we keep coming back to the connection between genes and information, between the chemistry of life and the logic of DNA. Computers nowadays are not merely useful analogies for describing the action of genes: computers themselves have become essential in analysing how genes work.

To have two technological revolutions arriving at the same time may seem a curious catenation of circumstances. But it is in fact a reflection of a slow change in the direction of scientific research, and probably also of the larger society in which it is carried out. We place an increasing reliance on information, and find that we are limited in what we can do not so much by the available power, or the materials to hand, as by the lack of information to tell us what to do next. This is seen in the enormous growth of 'artificial intelligence' as a science and not simply as a branch of science fiction, in the increasing reliance by manufacturers on computer simulations rather than test models to tell them if something will work, in our dependence on databanks and bureaucratic offices, and even in the inevitable form asking you to fill in your National Insurance number. Cecil B. DeMille proverbially advertised a cast of thousands in his spectaculars; now computer-guided special effects are the big draw at the box office. The executive status symbol of the 1950s was the bigger automobile; that of the 1980s is a more powerful desk computer. Even the weapons of global holocaust are changing from H-bombs which obliterate cities to much smaller cruise missiles with maps and computers to read them.

The new technology of genetic engineering fits perfectly with this mood. It is an extension of biology beyond the level of chemical reactions and bone lengths into the realm of control circuits and information theory. These esoteric subjects were studied theoretically by a few biologists in the last century, but it is only in the last two decades that substantial effort has been expended to find out if their theories are actually right in practice. This strange new biology will undoubtedly be a major science in the future, and as such it is bound to come back to the basic tool of the information age: the computer. Indeed, it is likely that the timeliness of genetic engineering, the way that it fitted in so well with our new, computer-orientated way of looking at things, contributed as much to the early enthusiasm for the subject as did its rather limited success in actual experiments in the late 1970s. After all, something so neat just had to work.

When molecular biology did turn to the computer, both parties were ill-prepared for the meeting. It has taken some years to adapt the programs, developed in 'artificial intelligence' research for problems as disparate as playing chess and making management decisions, to the examination of biological data. Some researchers still prefer to do their analyses 'by hand'. However, they are probably missing a lot. In Chapter 4 we mentioned the proto-oncogenes, which could be one of the basic causes of cancer. The discovery that one of these proto-oncogenes was related to a gene for a growth factor was made by computer comparison of DNA base sequences, not by laboratory experiment. Some of the controlling segments of DNA we have described, such as promoters, have also been found by computer studies initially, their existence being confirmed by experiment later. One such DNA sequence has been suggested as a controller gene which activates genes in the brain. If this computer-generated result turns out to be correct, it will be a start in the analysis of one of the most complex systems in all biology.

Probably the most productive area of computer applications today is in the study of evolution. By using computers to compare genes from such diverse organisms as wheat, humans, mushrooms and E. coli, scientists can probe the early history of life on earth 3,000 million years ago. Although these studies are of no conceivable direct use, they are great puzzles for students to sharpen their wits on and are the testing grounds for computer methods which industrial research laboratories will use later. The methods being developed today to tackle the knotty problem of the nature of ancestors of bacteria living in the hot springs in Yellowstone Park will be used tomorrow to predict the best promoter to use in a commercial genetic engineering project.

We have strayed a little from our path to glance at the more academic side of recombinant DNA, the side which generates the novel techniques that will be used by industry. When talking about research being performed today, this aspect is not too important. When deciding what might be possible tomorrow, it becomes central to our theme.

The study of genes has told us a lot about how they work. Has it told us enough to be able to manipulate them?

Maybe. Using a variety of methods, scientists can put their genetically engineered DNA into the cells of animals and plants: we will discuss these methods as they apply to mammalian cells in just a few paragraphs.

Wherever the cells came from originally, mouse or wheat plant, they must usually be isolated in a test-tube when we insert our new DNA into them. Thus we could do with these cells what we have already done with *E. coli* – we could insert a gene into them and then use that gene to make a protein. We could, for example, take the gene for insulin, hook it up to a suitable pro- moter (the promoter from the Lac genes, which Genentech used when they made an *E. coli* to produce insulin, would not work in animal or plant cells), splice this into a suitable vector and put the result into mouse cells. The cells could then be grown in a test-tube and used to produce insulin. But why do this when we have already done it with *E. coli*? Mammalian cells are much harder to grow than *E. coli*; they multiply more slowly and need expensive materials to keep them alive, and they must be kept in a rigorously sterile environment as otherwise bacteria would 'infect' them and kill them off. So although we could use mouse or dog or human cells to make insulin, it is easier and cheaper to use *E. coli*.

This is not true of all proteins, however. When we outlined how proteins are made we mentioned that the completed chain of amino acids could be altered after it had been assembled by one of the processes collectively called 'post-translational modi- fication'. While the basic theme of the process of protein synthesis is the same in all living things, the post-translational modifications are more specific to certain types of organism and are not shared so widely. We mentioned this in Chapter 8, and described how to get round this problem, genetic engineers can use yeast cells in place of *E. coli* for their manipulations. Despite being single cells which grow almost as fast as *E. coli*, yeasts are actually much more closely related to us than to *E. coli*. So their mechanisms for making post-translational modifications are more likely to be compatible with ours, and consequently the proteins produced by their cells are more likely to resemble the ones made by our own cells than are the products of a genetically engineered *E. coli*.

Another benefit of using yeast is that its DNA, unlike that of *E. coli*, is arranged like our own in a number of long, straight chromosomes wrapped up in histone proteins in a nucleus. Also, yeast's genes can contain the 'commercial breaks' which, as we mentioned, can 'interrupt' our genes. This has two important consequences.

Firstly, it is very hard to persuade *E. coli* to take up large genes, and impossible to make that organism decode the information in them if they contain the 'commercial breaks' present in our own genes. As many of the genes we would like to clone from mammalian DNA are both very large and contain these 'breaks', this can be something of a problem. However, yeast cells do not suffer either of these disabilities and so could happily cope with, say, one of the human genes for collagen which are over 50,000 bases long and contain dozens of gaps.

The second point brings us back to research again. Animals are very poor subjects for genetic research. They need lots of care and attention, they only breed infrequently (even fruit flies take several days to complete a generation), and they are big. While a geneticist might plan an experiment involving a thousand million *E. coli*, to undertake the analogous experiment on wheat or sheep, or even fruit flies, is clearly impractical. This is one of the reasons why the genes of *E. coli* are much better understood than those of wheat or sheep. Nevertheless, we *can* design a sensible experiment involving a thousand million yeast cells and so obtain detailed information on how yeast genes work, which is a step in the right direction. We can also put a sheep or wheat gene into a yeast cell and then carry out huge genetic experiments on the resulting recombinant yeast without covering the whole of North America with wheat or sheep.

And, of course, yeast is the obvious organism to choose to engineer if we plan to alter a process already carried out by yeast. Brewing and baking are manufactures which rely on yeast to turn a raw material into a product. If we wanted to use recombinant DNA techniques to improve baking, it would be more sensible to alter the yeasts already involved in the process than try to genetically engineer an *E. coli* for the complicated task of turning dough into bread.

Other factors can also influence a genetic engineer towards the use of yeasts. The way in which yeast genes work is very well understood; this gives the genetic engineer a better idea of what will happen if a new gene is put into a yeast cell, and hence how an experiment should be designed to make a gene do what it is supposed to do first time. Also, yeast cells have far less DNA than human cells − only 4.6 times as much as *E. coli* and 150 times less than man − which makes the technical hitches of gene cloning less troublesome.

However, this is just an extension of the sort of project we discussed in the last chapter. We are putting a new gene into a relatively simple, single-celled organism and then using that organism to make a protein for us. While yeast offers technical advantages over *E. coli* for some experiments and technical disadvantages for others, we are not forging new routes here, only extending old ones. That is not to belittle the work of yeast geneticists who have made yeast into an important industrial tool, but we started this chapter with the intention of leaving industrial microbiology and investigating the possibilities of genetic engineering in higher organisms.

The problem with genetic engineering higher organisms is one of complexity. We mentioned earlier that we should like to integrate DNA into the cell's chromosomes, and at the right place, too, although we do not know exactly what the right place is. Then there are the considerations of what DNA to put in, which promoters, enhancers, and so on, that the experimenter considers useful (or that the experimenter knows about!). The details are more complicated if we want our gene to be active only in a few specific cells, for example to produce the protein only in blood cells and not in the liver or brain. So far only a few of these problems have been sorted out.

There are now several ways of putting cloned DNA into an animal or plant cell. Plants have turned out to be somewhat easier than animals, for two reasons. Firstly, scientists have enlisted the aid of a bacterium called *Agrobacterium tumefaciens*. This bug, which causes crown gall disease in some plants, contains a plasmid which is partly responsible for its pathogenic effects. When it infects a plant the bacterium injects a few cells with this plasmid, part of which becomes integrated into the plant's own DNA. The cells in which this happens are transformed into a cancer-like growth, the gall. Thus from having a gene in a plasmid in a bacterium, the result is that the same gene is integrated into the chromosomes of a plant cell. In principle, then, all a genetic engineer has to do is splice the gene into the correct site in the *Agrobacterium tumefaciens* plasmid and let the bacterium itself take care of the tricky part of putting that gene into a plant cell. With some tinkering with the plasmid and some clever genetics, this sort of genetic engineering has now been achieved.

The other reason that plants are favoured is that they are easy to clone. We are back to cloning again, except that here we are talking about organisms each with billions of cells, rather than organisms each with just one cell, like *E. coli*. The cells from some plant tissues can be separated from each other and spread on a special material rather like jelly, which contains a rich mixture of amino acids, bases, hormones and other ingredients. The cells can grow on this, and form little heaps that look rather like a small fungus. If one of these heaps is then transferred to a new type of jelly containing different hormones, the cells in the heap will start to reorganize themselves into roots, stem and leaves. After culture in such a medium a complete new plant appears, and can be removed from the jelly and planted in normal soil. This new plant derived from one cell heap, which was itself derived from a single cell. In this way a particular plant can be broken up into many cells, each of which can be grown to make a new plant genetically identical to the original. This is a more effective way of producing a number of identical plants than breeding, which does not produce identical offspring, or taking cuttings (another, more traditional method of cloning), which produces identical plants but only produces a few dozen per parent.

Using the technique of plant cloning, scientists can genetically engineer, say, 1,000 plant cells, pick the one in which their manipulations have worked and grow an entire plant from it. Using the same technique, that plant can then be cloned to produce a hundred or a hundred thousand identical copies of it. Without this technique the scientists would need a much higher success rate in their initial genetic engineering, which – given the hit-or-miss state of the art today – would be difficult to achieve in most projects.

The cells which the genetic engineer is to clone in this manner to produce a new plant could be derived from a crown gall – cells infected by the *Agrobacterium tumefaciens* plasmid. In this way all the 'offspring' plants of the cloning procedure would inevitably have the plasmid DNA in them, and hence a good chance of having any DNA which we engineered into the plasmid as well. But we can side-step the use of the *Agrobacterium tumefaciens* plasmid as there are also methods of putting DNA directly into cells. For this, the cells must be isolated from their parent organism (imagine trying to inject DNA into a cell through six inches of bark). Then the scientist can either treat the cell chemically with the DNA – that is, just spread the latter over it and hope that the cell will take it up – or inject it with a very small needle. The first method is rather inefficient, but if a large number of cells are used in the experiment and the few which have taken up the DNA are then isolated, it is acceptable. The latter is quite efficient, but making a needle which will fit into the nucleus of a cell only a few millionths of an inch across is not easy. Both these approaches have been made to work on animal and on plant cells. In a fairly recent British experiment, a Cambridge group successfully got such engineered cells back into mice, where they mingled with the un-engineered cells to become normal mouse tissue. Some became germ cells (sperm), and so the 'new' gene was passed on to the descendants of the mice.

A third approach, and the most promising for the genetic engineering of animals (at the moment, anyway), is to use a virus. I mentioned in Chapter 7 that viruses could be used as vectors to clone DNA, but they have rather taken a back seat in our considerations since. This is partly because most viruses attack cells fast, destroying them before moving on, and this is not very useful if we want the cell to make a protein for several months to come. However, I pointed out in Chapter 4 that some viruses behave differently, lingering in our cells for months or years. *Papilloma* viruses are one such type. Another type, called retroviruses, actually go to ground in their target's DNA. They splice their own DNA into the target cell's chromosomes – they integrate themselves – and thus end up looking just like another gene to the cell itself. It is only when some signal tells them to awaken, a signal we still do not understand, that they become active. These viruses have been the subject of an enormous amount of research because they have been implicated as a possible cause of cancer. The HIV family of viruses, of which the AIDS virus is a member, are retroviruses, and some of them cause rare types of cancer in human beings.

It would be reasonable to think that any virus so tricky and potentially dangerous as a retrovirus would be of no conceivable use for the genetic engineering of humans. However, scientists are hoping to use retroviruses as vectors for genetic engineering in mammals. They have identified those parts of the virus's DNA that are involved with cancer, the parts which are related to the virus's own reproduction, and the parts which are connected with its integration into the chromosome. The latter parts are being used to make a vector into which DNA can be cloned, to be carried into the chromosome by the virus itself. The method is rather complicated, because if you have removed the parts of a virus that make it grow, how is it then to grow so that the DNA from it can be used to engineer? The answer is to use another virus to help, but then *this* virus has to be removed. However, some preliminary trials have shown that this is a very promising approach – scientists have successfully used vectors made from retrovirus DNA to put DNA into mouse chromosomes. The drawbacks have been illustrated in similar experiments, however: a severe mutation of a mouse collagen gene, causing a disease similar to the human disease osteogenesis imperfecta (see page 31), has been caused by a retrovirus DNA integrating itself into the middle of a collagen gene. What is to prevent our retrovirus-derived vector doing exactly the same thing in mice or humans? So far, nothing. Nor is there anything to stop that gene integrating next to a proto-oncogene and mutating it so that it becomes an oncogene, a gene causing cancer. Although a start was made on the task of targeting such vectors to specific parts of a chromosome in 1986, more work will be needed before we can say that such unwanted side-effects are unlikely to occur. And then, of course, we will have to face the horrendous complexities of the control circuits which regulate our genes. So although the use of retroviruses is a major step towards the goal of using genetic engineering to alter animals, it is only one step in a considerable hike.

On that cautious note, let us conclude our catalogue of difficulties in the genetic engineering of animals and plants, and have a look at some of the things which have actually been achieved.

Chapter 11
Moving towards Man

In 1980 a researcher at the University of California at Los Angeles (UCLA) claimed that he had made the leap from *E. coli* to man. In a single step Martin Cline, a research haematologist, had taken the techniques of recombinant DNA technology and had attempted to use them to cure human disease. In doing so he dodged all the regulations, ignored his colleagues' advice and brushed aside the considerable ignorance of the day concerning what such genetic engineering might involve. In the furore that followed Cline was reprimanded by his university, the National Institute of Health, and a horde of media 'critics' who held him up as an example of what scientists would do if not prevented by law. Cline lost a research post and a major research grant, and became an instant byword for tinkering where angels fear to toy.

The irony is that within five years of Cline's ill-fated attempt a dozen laboratories were undertaking preliminary experiments on the basis of the same ideas, but on a somewhat sounder factual footing; they have already engineered the genes of mice carrying a defective gene which causes disease, as a preliminary to trying the same thing on humans. Within a couple of years the barriers preventing genetic engineering of humans will have become as much legal and moral as technical, and those barriers will be the harder to sort out because of the confusion raised by Cline's experiment.

The idea that Cline was pursuing is worth examining, as it shows up some of the problems inherent in trying to transfer a technology from *E. coli* to man. We have examined the academic study of these differences in the last chapter – Cline's experiment was to run across them in practical form. Cline, as a haematologist, was trying to find a cure for a genetic disorder of the blood. The disease was thalassemia, which we met in Chapter 5; it is characterized by a lack of the protein haemoglobin in red blood cells, a deficiency caused by the absence of a single gene. Now what could be a simpler target for a genetic engineer? He need only put a working copy of the gene into the patient's blood cells, a copy cloned from a healthy person, and the patient would be able to make their own haemoglobin again. The disease would have been cured.

The various genes for haemoglobin (there are at least ten in man, producing variants of the different parts of the molecule) were among the first human genes to be cloned, and by this time had been intensively investigated. Cline therefore had access to the relevant gene. He tried a pilot experiment on mice; it had some success. So he applied for permission to go ahead with a human thalassemia patient. Permission was refused: the opinion of the molecular biologists was that the time was not yet ripe. In ten years, maybe . . . Cline said afterwards that seeing patients dying of thalassemia every day did not encourage him to wait ten years, so he side-stepped the ruling of the US medical establishment by going abroad. There are many times more thalassemia patients in the countries of the Eastern Mediterranean than in the United States, and so when Cline approached medical authorities in Israel they were more receptive than those at UCLA. (It is not clear whether he explained exactly what he planned to do; any lack of precision may also have helped their acceptance of the plan.) Cline obtained permission for his human experiment, and went ahead.

He first extracted bone marrow cells from two thalassemia patients. Red blood cells themselves do not divide: they are unique among the cells of the body in that they have no nuclei, no DNA. Thus they are running 'on automatic', with no instructions written in DNA to tell them how to adjust to changes in their environment or how to repair themselves. Consequently they wear out after a few weeks in the blood, and have to be replaced. The source of these replacement cells is bone marrow: the relevant bone marrow cells are called 'stem cells', as they are the 'stem' or 'rootstock' from which subsequent generations of red blood cells are produced. So if we want to put the product of a new gene into red blood cells and have it work there for more than a few weeks, we must alter these stem cells and not the red blood cells themselves. Thus Cline treated the bone marrow cells with one of the genes for haemoglobin and then injected the cells back into the patient's bloodstream, whence they would find their own way back to the bone marrow. He hoped that there would be sufficient stem cells in his bone marrow sample, and that enough of them had taken up the haemoglobin DNA, to make a difference.

Nothing happened. The bold, innovative experiment was a total washout. Some cynics suggested that if it had been a success Cline would have got a Nobel Prize and not a rap across the knuckles, but in this they ignore that success was very unlikely. Even at the time, it should have been clear that it had little chance of succeeding. Now we know more about how human genes function, thanks to research using those same recombinant DNA techniques, the chance seems even more remote.

Let us go through the experiment again to see why this is so. Extracting bone marrow cells and putting them back into a patient is an established technique. (It is also excruciatingly painful, even with anaesthetics.) So our starting material is on firm ground. Then we want to put genes into these cells. This raises two questions: Which genes, and which cells?

The haemoglobin genes are all quite large, several thousands of bases long. They occur in two clusters, each some tens of thousands of bases long. And, as we mentioned in Chapter 10, each gene may be influenced by its general surroundings in the chromosome, which translates in DNA terms into anything up to 100,000 bases. So which gene was Cline to use? The section of DNA which held the information to make the haemoglobin protein? The whole cluster of genes, just in case there was one of those enhancers (not actually discovered at the time, although the existence of such pieces of DNA was suspected) lurking among the genes? Or all the 100,000 DNA bases flanking the haemoglobin gene, which almost certainly held all the pieces of DNA needed to make the haemoglobin gene work properly, but which might also hold other genes that Cline did not want (and which anyway would be so big that, if pulled straight, would be ten times as long as the bone marrow cells)? As only the DNA for the first of these options was available to Cline, he decided to go ahead with that. Almost certainly this was a mistake.

Then, to deal with our second question, he had to choose which cells to treat. He already knew that he had to get the haemoglobin gene into the stem cells in the bone marrow, but no one knows how to identify these among all the other cells in the marrow itself. He chose to treat *all* the cells in the bone marrow. In the event, either there were too few stem cells in his sample, or (more likely) he managed to get the recombinant DNA into too few of them. The techniques for putting DNA into human cells were quite inefficient, a state of affairs that the new vectors made from the DNA of retroviruses is meant to improve. At the time, though, getting one in 10,000 cells to take in a piece of the DNA was the best possible result that could be achieved. Cline could not therefore have expected more than one in 10,000 of the stem cells in the bone marrow sample to have received the new gene, and that expectation would have been optimistic. This sample was certainly no more than 10 per cent of all the stem cells in the patient's body; thus if every one of the 'new' haemoglobin genes worked properly, only about ten stem cells in a million would have produced haemoglobin. The stem cells in thalassemia patients can usually do better than that unaided.

How far have we come towards making a successful repeat of this sort of experiment? The most spectacular results have come from the laboratories of Richard Palmiter and Ralph Brinster in Washington and Pennsylvania respectively. They have performed several experiments that have overcome or side-stepped the problems that Martin Cline encountered. Their first experiment gave mice the gene for human growth hormone.

They made two improvements on Cline's method. First, adult mice were not the target for this genetic engineering. Instead, DNA was injected into a fertilized mouse egg, which was then implanted in a mouse mother and allowed to grow into a baby mouse. Thus all the cells in the baby mouse received the 'new' gene, as they all derived from the engineered, fertilized egg, and if the genetically engineered mice later mate they will pass on the 'new' gene to their offspring, just as they would any other gene. This is a realization of the project we suggested in a fanciful way in Chapter 2, when we suggested making a family of blue mice by genetic engineering. In this case the gene was not the imaginary one for blue hair, but the real one for growth hormone.

Another useful advantage of this method is that it is very efficient. The DNA is physically injected into a mouse egg, so we can be sure that the DNA will get into the egg's nucleus. This contrasts with the methods used by Martin Cline and by most other laboratories, which are inefficient in putting DNA into any cells and rely on separating out the genetically engineered cells after the genetic engineering step has been performed. We have already seen the latter process carried out in exactly the same way on *E. coli*, to discover which of a huge collection of cells contain a genetically engineered plasmid. If the plasmid contained the gene for, say, resistance to penicillin, then we can simply grow the *E. coli* in the presence of penicillin and any cells which have not received the plasmid will die. The same sort of thing can be applied to human cells: we can give them a gene which confers resistance to a drug that is normally poisonous to human cells, and then, after our genetic engineering, expose all the cells to that drug. Only those which have received the 'new' gene will survive. Cline could not do this to his human patients, of course, as all the other cells in the patients would also be susceptible to this poisonous drug. The genetically engineered stem cells would indeed have survived and replaced all the un-engineered stem cells. But unfortunately all the other cells in the body would have died off. While not of much use in treating whole people, such a method is nonetheless very useful for raising the efficiency of genetic engineering performed on isolated cells in the test-tube.

If you can separate genetically engineered cells from their more numerous unaltered sisters, then another approach becomes feasible. This is to alter an animal cell in the test-tube, and then inject it into a growing embryo. If this is done just a few days after conception when the embryo is made up of only a few cells, the new cells you are adding will make up a substantial fraction of the resulting mouse. Usually, injecting cells into an embryo is not a good thing to do: the cells will have acquired some of the characteristics of cancer cells during growth outside their normal body, and so will either kill the growing embryo or will be killed by it. However, not all cells are so lethal; in particular, the cells from a rare type of cancer called 'teratocarcinoma' are capable of joining with those of an embryo and becoming part of it with no adverse effects. If we genetically engineered such cells from a mouse and then injected them into a mouse embryo, their descendants would end up in all the tissues of the animal's body.

Palmiter and Brinster used the first approach – that of injecting fertilized eggs with their recombinant DNA. Their other innovation was to hook up a promoter to the growth hormone gene, so that they could be certain that any recombinant DNA they got into the mice would be active in production of growth hormone protein. The promoter they used was from a mouse gene, so that they were fairly sure that it would work in mouse cells.

The experimenters used several sophisticated methods for detecting how active the 'new' gene was in the genetically engineered mice which resulted from this experiment, but the most dramatic demonstration was the simplest. The genetically engineered mice were up to twice the size of their normal littermates. Their bloodstreams carried far more than the normal amount of growth hormone, so they grew faster and in the end were bigger. It was a vivid demonstration that genetic engineering of a fairly simple sort *could* alter an animal as complex as a mouse.

The news was not quite so good when they tried to use this method to cure a genetic disease. An inherited disease from which mice can suffer is similar to some sorts of hereditary dwarfism in man: both are caused by lack of growth hormone. So there was great interest in the possibility that this method of placing recombinant DNA in mice could be used to correct the deficiency, because that would be a good indication that a similar approach might work on sufferers from the human disease. The fertilized embryos whose parent mice were dwarfs were injected with the growth hormone gene, but alas! the embryos still grew to be small mice, although not as small as their unaltered littermates. They also developed other problems, such as partial sterility, from which their littermates did not suffer. So although a significant step forward has been made, this technology is still not sufficiently advanced to give complete success even on such simple trials as this.

Interestingly, when the same experiments were repeated with the growth hormone gene alone, with no extra promoter added to it, they were a complete failure. The dwarf mice remained dwarf. Although the growth hormone gene must have its own promoter, it does not seem to work properly in this experiment. Maybe a piece was accidentally chopped off during the gene cloning phase, or perhaps the recombinant DNA integrated at the wrong place in the chromosome to be able to work on its own. The same disappointingly negative results have come out of a number of other experiments in which recombinant DNA has been injected into fertilized eggs (frogs' eggs are a particular favourite for this purpose, as they are very much larger than mouse eggs). However, the picture is not all gloomy: recently an increasing number of scientists have succeeded in putting recombinant DNA into isolated cells or into entire animals (usually mice, although rabbits, pigs and others have been used) and making the recombinant DNA work more or less as it should.

So there is little doubt that the problems of making recombinant DNA work for us in the cells of mammals will be overcome, just as the problems of making recombinant DNA work in *E. coli* were overcome. Seven years ago when Martin Cline did his experiment on thalassemia patients, the experts poo-pooed the genetic engineering of mammals as science fiction, and not something that would be feasible for a decade at least. Now such techniques are on the boundaries of fact (although researchers now prefer to call these methods 'gene therapy' to disguise the connection with bad old science-fictional 'genetic engineering'). If these problems can be ironed out, how far shall we be from a true cure for *human* genetic disease, like growth hormone deficiency?

The answer is: rather further than you might think. There are major differences between curing most human genetic disease and making giant mice. The mice were engineered soon after conception, when all their genetic potential was held in one cell, the egg. But only a few human genetic diseases can be diagnosed before birth, and then no earlier than at eight to ten weeks' gestation. By this time a growing embryo will contain billions of cells, and cannot be removed from the womb to cure. So by the time tests can be carried out to see if an embryo is going to be born with thalassemia, it is too late to give it 'new' genes to correct the disease. The only way of reproducing Palmiter and Brinster's experiments with human subjects would be to treat all human eggs which might have some risk of the disease just after they are fertilized, and then put them into their mother's womb. This might be fine for the eggs which were going to grow up to suffer from the disease – they might be cured; but the others which were destined to grow up perfectly normal would have a lot of unnecessary recombinant DNA loaded into them, and that DNA would almost certainly cause some side-effects in its own right. This does not seem a very promising approach at the moment. The way round the problem is to be able to diagnose these genetic diseases in the fertilized egg, but how this would be done is quite unknown with today's technology.

The other route is to try to treat the affected infant after it has been born – this means at any time from immediately after birth to adulthood, as any method which can be used to treat an infant should also be applicable to an adult. Two methods show promise, although both have rather drastic drawbacks at the moment.

The first is to use the retroviruses we described at the end of the last chapter. These viruses naturally integrate themselves, as part of their life-cycle, into the DNA of the cells that they infect. As we mentioned, we can use this ability to transfer a gene into a cell's chromosomes with high efficiency, and so we could, in principle, engineer a large percentage of the blood cells of a patient suffering from thalassemia by treating them with a retrovirus into which we have spliced the gene for haemoglobin.

The drawback is that these are dangerous viruses, not merely because of the disease they cause but also because of this very ability to join into our chromosomes. This could in itself cause a new genetic disease, or could initiate a cancer by altering the function of a proto-oncogene. It makes no difference that such an event may be unlikely. If we were treating all the bone marrow cells in a thalassemia patient with such a piece of recombinant DNA, then we would need only one cell in a hundred million to be altered to a cancerous state to give a patient leukemia just when a cure of thalassemia had been achieved. Until these side-effects are tamed, retroviruses will remain a laboratory tool which is not suitable for human use.

Having said that, evidence has emerged in the last year that retroviruses, and maybe other recombinant DNAs, can be targeted to specific, harmless areas of the chromosome. So this type of treatment may be nearer than we thought in 1985.

The second drawback of this approach is that it can only work for a very few genes. A growth hormone gene can work in almost any cell. The growth hormone protein will be made by the cell and, because it has a signal peptide attached (Chapter 3), it will immediately be pumped out of the cell into the bloodstream. Once in the blood, it will circulate to wherever it is needed. However, most genes are not so cosmopolitan. As we mentioned above, the genes for haemoglobin need to be active in the stem cells which give rise to red blood cells. It would be worse than useless for a retrovirus to carry a 'new' haemoglobin gene to every cell in a thalassemia patient's body, for then the kidneys, brain, bones and many other tissues would make this protein, possibly with as injurious an effect as the original thalassemia. Even the retroviruses which did get through to the stem cells would not necessarily be useful, as the haemoglobin genes have to work in accurate unison to produce a working red blood cell. If the 'new' gene produces too little haemoglobin, or too much, that too could be as bad for the patient as the original thalassemia. For all these reasons, such a blanket approach may not be generally very useful.

An alternative to using retroviruses as a method for carrying DNA into cells was tried out in France towards the end of 1983. The French group took the DNA which contained the human insulin gene and wrapped it up in the fats and proteins which make up the plasma membrane of red blood cells. The plasma membrane is the normal outer covering of all human cells, which recognize each other by this common feature. Thus to other cells these little 'packets' of DNA looked not like DNA but like small red blood cells. These packets were injected into rats, where they made their way around the bloodstream to the liver. The liver normally removes old and battered blood cells from the circulation, and the packets were removed promptly and taken up into the liver cells. Some of the DNA inside them finished its journey in the liver cells' nuclei, and in this way the living rats ended up with a human insulin gene in them. The DNA produced insulin, too, for a few hours before it was destroyed by the liver cells as an unwanted foreign molecule.

This approach is a valuable one as it allows us to target a piece of recombinant DNA to a specific tissue, in this case the liver. By using other plasma membranes for wrapping the DNA, and antibodies to target those membranes to different sites, it may be possible to insert DNA into many of the tissues of the body (the brain is the most notable exception). In this trial experiment the effect was short-lived, but this too might be overcome.

Unfortunately this approach does not get round the problem of how to make sure that the integration of this DNA into the chromosome does not itself cause genetic disease or cancer.

To sum up: we can insert DNA into cells in a test-tube fairly easily, but getting it into intact human beings is much harder, partly because of the potential problems the DNA itself could cause and partly because we do not have any good method of doing it anyway. Once these difficulties have been overcome and the various controller genes, promoters and so on which regulate our genes are better understood, it should be possible to treat a range of genetic diseases resulting from a defect in a gene that is active in adulthood, for example thalassemia. Even then, such diseases as muscular dystrophy and cystic fibrosis will still be rather hard to attack, as at least some of their damage is done while their victim is still in the womb; how to treat a baby *in utero* without harming the mother is a question for which the technology of today offers no answers.

Two classes of genetic disorder will be amenable to treatment by recombinant DNA methods in the near future. The first is the type of disease which can be alleviated by giving the patient a known protein. (Many genetic diseases may be treatable by giving patients a protein, but as yet we do not know, in most cases, which protein should be used.) In practice this means injecting a protein into the blood, and this limits us rather severely as most proteins have to work inside cells, not in the bloodstream outside them. However, there are a few important proteins which do work in the blood. One which we have already met is insulin, which must be in the blood in order to be carried round to the different cells of the body. Another group of examples are the enzymes responsible for making blood clot. If a patient lacks one of these enzymes, then we would only have to inject that protein back into the patient's blood to correct the defect. This is not true genetic engineering, of course – the protein will not last long in the blood and will have to be injected again when it runs out. But we can use recombinant DNA to make the protein. An example of just such a genetic disease which will be treated in this way in the very near future is haemophilia, the sex-linked genetic disease (see page 60) caused by a lack of the protein Factor VIII. An *E. coli* producing Factor VIII has been engineered, and should soon be ready to produce the protein for commercial use.

The second class of treatable genetic disorders are those where our limited knowledge of 'gene therapy' can actually be used directly to cure the disease. The victim in the case of these inherited conditions lacks an enzyme in the blood cells, or in the bloodstream around those cells. Fortunately, this is a very rare class of disease indeed: although there are dozens of enzyme deficiencies of this sort, on average each one affects only a handful of people in every million. An example is phenylketonuria (usually abbreviated to PKU), a rare disease caused by the lack of an enzyme which disposes of surplus amounts of one of the amino acids. Every child in Britain is tested at birth for PKU, not because it is a common disease (it affects about one in 20,000 babies) but because the test is extremely cheap and the consequences, if it remains undetected, are very severe. Its symptoms – severe mental retardation – arise because of the accumulation of poisonous chemicals in the blood which the liver cannot break up: usually the missing enzyme would take care of them. The obvious treatment for the disease is to prevent those chemicals from ever being made. This could be done by modifying the diet so that the victim makes far less of the poisons in the first place (this is the treatment at the moment), or by giving back to the victim the missing enzyme. In fact, getting any enzyme into the PKU victim would be of benefit, and it would not matter too much where it ended up as the poisons could be broken down anywhere in the body.

In this case we can consider repeating Martin Cline's experiment, with one major difference. Now we can take bone marrow cells, give them the gene for the missing enzyme, put them back into the patient, and expect to see some result. We would not need a retrovirus vector molecule because it would not matter if only a few cells received the 'new' gene. Any cells producing any enzyme will be of benefit to the patient. We now know the promoter needed to make it work efficiently, and can do some preliminary screens on our engineered cells to see if any of them have acquired cancerous properties. We are aiming to get our gene into any cell which will end up in the bloodstream, not just the very specific stem cells which are the originators of red blood cells and which were Cline's target. Thus we know which gene to use, what cells to use it on and how to use it. Have we overcome all the problems of Cline's experiment?

Several scientists believe that we have. They intend to try this type of 'gene therapy' on animals and then, soon, on humans. PKU will not be the target, as the disease is easily treated by less drastic methods. The two diseases at the top of the list of candidates are Lesch-Nyhan Syndrome (a deficiency in an enzyme which synthesizes one of the bases in DNA, causing severe mental retardation and self-destructive behaviour), and AMA deficiency (which causes a general failure of the immune system, rather as the AIDS virus does). Unless there is yet another unforeseen hitch, this should be a major step towards the cure, as opposed to the alleviation, of genetic disease. Unfortunately, the vast majority of sufferers from inherited diseases will have to wait a few more years before their afflictions are amenable to this approach. To say that the cure of one genetic disease is a herald for the imminent cure of all genetic diseases is to ignore the fact that the disorders themselves are as varied as the people they afflict.

So far in this chapter we have talked exclusively about the medical applications of the genetic engineering of higher animals. This exactly parallels the earlier successes of genetic engineering of E. coli. The highly publicized, highly profitable biomedical applications came first, and only later did scientists turn to the production of food and chemicals. Will the same shift in emphasis occur in the application of recombinant DNA technology to higher organisms?

I am sure it will, but there is a question concerning this shift: What will the scientists choose to do? Many obvious manipulations come to mind, but on closer examination they often turn out to be economic non-starters, or even completely useless. For example, we could attempt to duplicate Palmiter and Brinster's experiments with mice, using a cow. The techniques for manipulating fertilized eggs of most major farm animals are well worked out, and so a simple combination of Palmiter and Brinster's methods with the technique of in vitro fertilization would give us the capability to create a giant cow. But would this beast actually be useful to the farmer? The amount of meat or milk produced per pound of fodder is unlikely to be very different from that produced by a normal cow. The ratio of meat to tendon and bone would probably go down because, as in the human disease acromegaly (see page 52), an excess of growth hormone would cause the bones to enlarge to a greater extent than the muscles. The cow would also be more prone to broken limbs, as are very tall humans, and might suffer other side-effects. It would also be too big for all the farmers' existing equipment. The only major gain would be that the animal would grow faster (and eat faster, too), saving on heating costs in sheds and possibly promoting a quicker turnover.

Despite these potential problems, in late 1984 several groups announced, or at least hinted, that they were planning to put a growth hormone gene into farm animals, and in 1985 rumours were abroad that a 'giant pig' had been created. Whether it will actually turn out to be a giant must wait a little while, of course: unlike mice, pigs do not grow to their full adult size in a few weeks. So great is the enthusiasm for the new technology that the practical use of this animal seems to have been rather played down. Instead of discussing this aspect, the battle lines of headline-writers have been drawn up along the same well-worn hilltops they occupied a decade ago when fighting the first round of public debate on genetic engineering. On one side, the experiment will open the way to unending supplies of super-cheap meat and enough food to feed the starving. On the other, new disasters which might arise from the widespread use of 'giant' animals are dreamed up (remember the giant ants which caused so much trouble in the horror film *Them*?). It is all rather similar to the unruly series of claims made for and against genetic engineering of bacteria to produce insulin and interferon, discussed in Chapter 8, and just as unlikely to lead to an informed debate.

There are some notes of caution to be sounded about the genetic engineering of farm animals and plants. For example, even if the 'giant pig' project does turn out to be economic, it is not certain that it will turn out to be safe. Clearly the risks are unknown, so we must rely on what analogies we can draw with other agricultural practices to see if there is any real chance of a danger to humans. One analogy is with the widespread use of antibiotics in animal feed. For obscure reasons, adding very small amounts of antibiotics to the food of pigs and cattle being fattened for slaughter makes them put on weight faster than other, untreated, stock. The doses are much smaller than those that would be given to cure a bacterial infection, and little remains in the slaughtered cow. What harm could this do to people who eat the meat? In the short term, no harm at all. However, in the long term it has been found that the treated animals were becoming hosts to bacteria which were resistant to antibiotics – a direct result of feeding them those antibiotics. The existence of such bacteria has been known for years and there had been rising concern that they might cause a health hazard to humans, but it was only in 1984 that several human deaths were shown to be the result of infection by antibiotic-resistant bacteria originating from cattle fed with antibiotics. Of course, growth hormone is not an antibiotic, so the same problem will not arise here. But it is a powerful hormone, and one whose long-term effects on adults are not well known simply because there had not been enough of it with which to experiment before the advent of recombinant DNA techniques. While growth hormone is essential for normal growth and is an almost miraculous cure for some cases of dwarfism (as antibiotics are an almost miraculous cure for infection), it is far from clear whether taking a dose of growth hormone every time we eat pork for the rest of our lives would be such a good idea. This is not a reason to forego the experiment (and even less a reason to ban all recombinant DNA research), especially because the problem is a technical one which should be answerable by precise technical means – a very different barrier from the legal ones which beset genetic engineering today. But it is a reason for caution.

Other, more speculative, projects can be imagined, but there are few simple improvements which can be made to domestic animals. We might want to breed sheep whose wool grows faster, or longer, or finer. There would be two approaches to such a project. The first would be to find a 'wool hormone' which made the sheep grow better wool; if we could find such a thing, then we could use recombinant DNA techniques to make it, or we could put the gene for the hormone into the sheep so that they made it themselves. However, this presupposes that we know such a hormone exists, which at the moment we do not. The first step in this kind of research project, then, would be to ask a veterinary scientist to investigate the existence of 'wool hormone'.

The other possibility is to engineer the sheep itself, altering the cells which manufacture the wool so that they behave more to our liking. But this is a far harder task, as it requires us to make a gene, or more likely a whole set of genes (which we have not yet identified), work in the right cells at the right time at the right levels (which itself requires us to understand exactly the function of all the controller genes which regulate those in the 'wool cells'). As we discussed in Chapter 10, we are far from understanding how any controller genes work in something as complicated as a sheep. We might want to make a cow produce low-cholesterol milk; but many genes provide the information to make the fats and cholesterol in cows' milk, and we would have to alter them all so that they worked properly in unison to make a usefully modified cow. Again, this is far beyond present technology.

This might seem a rather negative outlook, but it reflects our present ignorance about how genes operate, and the limitations of recombinant DNA technology. We can put a gene into a cell or a fertilized egg, or even (soon) into an adult animal. But to make the gene work requires an exact knowledge of the controller genes, the promoters, enhancers and all the other panoply of bits of DNA which make sure our own genes operate on time and on target. To make growth hormone is simple – we just need the gene to be 'on' in any cell it lands up in. But to make wool requires much more precision, as exactly the same gene could end up making hair or toenails or warts (all of which are composed largely of the same protein). In the same way, a noise that is just an uncontrolled racket is easy to make, but a regulated, rhythmic, meaningful noise like Brahms' third symphony (as Basil Fawlty would have it, 'Brahms' third racket') takes vastly more organization both in composition and execution. Giving animals the genes to make single proteins is fairly straightforward (one Japanese company is reputed to be developing a silkworm which makes interferon as well as silk), but altering the shape or behaviour of an animal is beyond us today. New technologies may change this by the year 2000; they will not by 1988. Once we get beyond the simplest of operations it is likely that many genes will have to be considered in our calculations. As we mentioned in Chapter 2, even such apparently simple characteristics as how tall we grow can be the result of the action of many genes. To alter such morphological features of animals, we will have to understand the action and control not merely of one gene but of dozens.

The prospect for the genetic engineering of plants is rather rosier, and has been making terrific progress in the last few years. Plant molecular biology was something of a poor relation to animal research, as it only be applied to such mundane pursuits as growing food, rather than to glamorous projects such as curing cancer or old age. However, the techniques for placing DNA in plant cells and then (just as important) growing the resulting cells into a clone of identical plants have been developed to such an extent that they are almost routine today, and consequently the methods have been applied to several projects.

Plant genetic engineering has two other significant advantages over animal engineering, apart from the technical ease with which it can be done. The first is also technical, in that plants are more tolerant of man's abuse than are animals. If you cut a leg off a cow, it will take some quick surgery and careful veterinary care to stop the cow from dying. But cutting a major branch off a tree will do little damage to the tree. The cow's leg cannot be encouraged to grow into another cow, but in many plants a severed branch can be quite easily encouraged to grow roots and become an independent plant. This is reflected in the ease with which individual cells from a plant can be encouraged to grow into entire new plants in the cloning procedure we mentioned in the last chapter. This means that genetic engineers need be less worried about their manipulations having lethal side-effects on the plant. A serious side-effect is likely to kill an animal, whereas a plant would be more likely to put up with it.

The second aspect is the number of commercially useful projects which can be performed with existing technology. There are two immediate areas of interest (a fact which contrasts with the relative lack of projects for the genetic engineering of farm animals): these cover the development of specific proteins to increase the food value of plants, and giving them genes which confer resistance to pests. Both rely on giving plants single genes for proteins that they themselves can produce in any cell – the only sort of genetic engineering we know how to carry out at the moment.

For an example of the first case, we could cite the maize plant. The proteins in maize flour are low in the amino acid lysine which humans need in their diet, and therefore maize flour is not an ideal source of protein if it is a person's principal source of nourishment. To correct this imbalance, we might engineer a maize plant to produce a protein which has a lot of lysine in it, for example a histone protein (some of these contain a great deal of lysine). The total protein content of rice flour could be increased simply by giving the rice plants a gene for an innocuous protein like haemoglobin, with a very powerful promoter attached, so forcing the cells to make a large amount of that protein.

A problem arises here – one we encountered in the genetic engineering of animals, too. As we mentioned in Chapter 3, plants have tissues just as animals do. The part of the maize cob in which the starch and protein are stored, and which is turned into flour, is a tissue. We want our 'new' protein to be made in this tissue, for two reasons. Firstly, this is the bit that is going to be made into human food. It is no use making the leaves of a maize plant very nutritious if we never eat them. There is a further reason why we would prefer the 'new' protein to be made only in the part of the plant which is to be used as food. The leaves and stems are made up of other types of tissues, ones which are not suited to storing large amounts of protein. These tissues may work rather badly if we load them up with our 'new' protein. In an animal, such poor functioning would be likely to be fatal; an engineered plant would almost certainly survive, but would be unable to put nearly as much of its energy into growing as the farmer would like. So although this sort of project could be made to work in as much as a 'new' protein could be made in a maize plant, it will be far harder to make this engineered maize plant a realistic competitor for existing strains.

Another sort of additional protein to improve food value could be an enzyme which would perform some function the plant cannot perform on its own. The prime example here lies in the ability to 'fix' nitrogen. All plants need the chemical element nitrogen as a basic food (as we do: we get ours by eating plants and other animals); most derive it from nitrates, chemicals which occur naturally in the soil. If those chemicals run out, the farmer must supply extra nitrates as fertilizer, or allow the field to lie fallow to regenerate its natural nitrate supply. Both are expensive things to do. However, some plants have a third course open to them: they can get nitrogen from the air. Eighty per cent of the air consists of nitrogen, but it is very hard to turn the nitrogen in the air into the nitrates which plants use. Some plants can take nitrogen from its free, unattached state in the air and 'fix' it into nitrate chemicals. In fact, the trick is done by a particular type of bacterium living in the plants' roots: in return for the nitrates which the bacteria give the plant, the latter provides the bacteria with sugars and other foods, and with an enclosed, protected place in which they can work. Clover is a plant which can use bacteria in this way, and this is why it is encouraged to grow on fallow fields: it is making those nitrates which will be ploughed back into the soil at the end of the season.

This seems a very valuable thing to be able to do, but few plants can do it. There are three ways in which a farmer might consider using the enzymes which 'fix' nitrogen to provide other plants with more nitrates. The first would be to use the bacteria to make the nitrates in an industrial process, as Genentech used bacteria to make insulin. This would not remove the need for nitrates, of course, but would be a way of making them. However, it takes a lot of energy to 'fix' nitrogen in this way, so much so that it is actually cheaper to use the normal chemical process instead, although this involves several chemical steps at high temperatures and pressures. Consequently this solution may not be so useful after all.

The second approach could be to give plants the genes to make those 'fixing' enzymes. This would seem straightforward, as only a few enzymes are involved and it does not matter where in the plant they are made. Unfortunately, this would not work either. The enzymes are extremely fussy about how they operate and they need special, cosseted surroundings if they are to work at all. This means that we would have to put several other genes into the plants in order to make the proteins for those surroundings, and then we would have to make sure that the proteins are all made in the right amounts at the right place and time – we are getting back to the realm of controller genes again. This is not feasible at the moment.

The third approach is the most promising, and is being actively pursued. Why do these talented bacteria only work in conjunction with a few specific plants, such as clover? Why not with wheat or maize? Because those special plants make a set of proteins which form a sort of 'docking mechanism' for the bacteria to latch on to when they first form their intimate relationship with the cells of the plant's roots. These proteins form a 'welcome mat' for the bacteria and tell them where they are wanted.

So we could give a wheat plant those 'welcome mat' proteins and see if the bacteria then latched on to its roots. If it did, we would end up with a wheat plant which needed no nitrate fertilizer because it could literally make its own out of thin air.

I have gone into this in some detail because it illustrates that an apparently simple problem like using enzymes to 'fix' nitrogen can have a multitude of solutions, and in the genetic engineering of plants it is often quite likely that one of them has a chance of working. There are other enzymes which we might want plants to have – to regulate how much sugar they make, how tall they grow, how fast they sprout, and so on. The farmer might be interested in manipulating any of these, and the plants might be sufficiently flexible to withstand the result.

So far these goals are medium-term ones; no one has yet actually produced a genetically engineered crop plant. Nevertheless they do not involve anything which cannot be done, either with present technology or with that technology augmented by knowledge which we know how to obtain. The actual results to date have been rather limited, but this is natural as plant genetic engineering has only existed for a few years. A major breakthrough came with the demonstration that genetic engineers could transfer a gene from one plant to another and make sure that it worked in its new home. The first such result was the transfer of one of the genes for a major protein in maize flour (which is short of lysine) to several other plants, notably tobacco. Just as Palmiter and Brinster's 'giant mice' were not meant to be a serious agricultural project, so this process is not for immediate farm use. It took a lot of trial and error to make this experiment work, and now that it has, more ambitious and more useful projects can be attempted.

(Why tobacco? we may ask. Not because maize flour makes good cigarettes. The tobacco industry invests some of its abundant profits in plant research, and naturally encourages researchers to work on tobacco plants in the hope that they may turn up something useful. This makes the tobacco plant an attractive research subject for botanists short of research funds. Thus tobacco was one of the first plants to be cloned from single cells, and those cells were some of the first cells to have foreign

DNA placed in them. This means that, for a genetic engineer contemplating a novel project, tobacco is an ideal subject because he has far less to find out about the cloning of the cells, and so can concentrate his efforts on manipulating the DNA. Later, when the project has worked on tobacco, he can move on to less well studied but more useful plants.)

The other major aim of plant genetic engineering is to increase pest resistance. Animals have a complex immune system to resist attack by parasites (see Chapter 4). To make that system more efficient by genetic engineering could be extremely difficult, because of all the dozens of types of cell and hundreds of types of protein whose actions dovetail together to make the whole thing work. But we do not need to; we use immunization instead. Plants do not have an immune system. Instead they fight off parasites, and animals which would eat them, with a wide variety of chemical weapons. The resins in pine bark and poison ivy, the poisons in hemlock, deadly nightshade and foxglove, the drugs quinine and opium – all these are examples of chemicals made by plants to fend off attackers. It is quite plausible, then, to take the genes for the enzymes which make these chemicals out of one plant and put them into another. One particular project is ready to be tested under field conditions, but has been delayed because of legal problems similar to those which prevented the testing of the 'frost-free' *Pseudomonas syringae*. A bacterium which lives in the soil called *Bacillus thuringiensis* produces a protein that is very poisonous to insects but is harmless to other animals, and to plants. Two plant genetic engineering research companies, Agrigenetics and Plant Genetic Systems, have transferred the gene for this protein into tobacco plants and in due course will be able to see if the resultant plants really are poisonous to insects and hence proof against insect attack. They are pretty sure that they will be, once they get the 'new' gene working in the plant, and they are also sure that the result would be safe for human consumption. The *Bacillus thuringiensis* protein is already used as an insecticide in the tropics. Monsanto are using the same gene in a slightly different way: they have put it into another bacterium and are planning to produce a spray made out of that new bacterium for use on the leaves of crop plants. Clearly the Monsanto option means that plants can be protected much sooner than those that are subjects of experiment by Agrigenetics or Plant Genetic Systems: once Monsanto have their engineered bacterium, it can immediately be sprayed on to any plant. However, the plants will need to be re-sprayed at regular intervals, while all the descendants of an engineered plant will produce the new protein of their own accord.

Several projects of this sort are on the drawing-board now, and Cetus, the first genetic engineering company to be formed, has filed an application to test such a resistant strain under field conditions. Interestingly, Cetus is keeping very quiet about exactly what the plant is, and what it is meant to be resistant to. This contrasts strongly with the razzamatazz which accompanied all the early successes of genetic engineering in the late 1970s. Genetic engineering has travelled the long road from academic discipline to industrial secret. In Chapter 13 we shall be looking at another version of the use of genetic engineering to give a crop resistance to attack, this time from a chemical, as it illustrates a darker side of the transition from academic study to industrial exploitation.

We should bear in mind that some of these projects are not really revolutionary in conception, unlike the methods being used to carry them out. Placing new genes in plants is the latest application of biotechnology, but improving the food quality of maize is not a new idea at all. Indeed, as we mentioned in Chapter 2, such has been the success of more traditional forms of genetics in precisely this area that the maize plant with which we are familiar has been created by selective breeding from quite different ancestors. Proponents of plant breeding are sceptical of the claims of genetic engineers, believing that they can achieve the same results by more traditional routes. In the improvement of the food quality of maize, they could well be right. However, breeding is limited by one crucial factor: it only happens between members of the same species. If we want to improve the quality of maize as a food, we can breed from the cobs which have the highest protein or calorie content, or grow the fastest, or have the best resistance to insect pests. We can improve the quantitative nature of the crop in this way, sometimes very dramatically, although this can take quite a long time. Here genetic engineering can only offer to reach a similar goal by a faster or surer route, and even then the complexities of performing the experiments, which we have touched on above, may mean that the genetic engineering route ends up by being no faster than any other. But if we want to give maize the ability to 'fix' nitrogen or to fight off pests with the *Bacillus thuringiensis* protein, then we must 'breed' it not with another maize plant but with a bacterium, and in normal breeding that is impossible. For such projects genetic engineering is not merely another way, it is the only way.

So genetic engineering of higher organisms holds high hopes in some limited areas, just as bacterial genetic engineering did ten years ago. These areas are far more limited than popularizers of the subject seem to imagine. The treatment of one human genetic disease by 'gene therapy' in the near future does not mean that all these diseases are about to be abolished. The experimental creation of 'giant pigs' does not imply that a complete genetic revolution is about to overtake farming. The problem is not that either of these broader goals is impossible in principle, but that our knowledge today is too limited to allow us to perform most of what we would like to do.

This is a very familiar situation. Exactly the same thing could have been said of bacterial genetic engineering in the early 1970s. The genetic engineering of higher organisms will follow a parallel course to that of bacteria. What is a laboratory curiosity today will rapidly be exploited in a variety of contexts in the next few years. What is now a technically intractable problem, to which we can nevertheless see an obvious solution and a way of reaching it, will also be attacked and solved; giving wheat those 'welcome mat' genes to allow it to offer a home to nitrogen 'fixing' bacteria is probably such a problem. We know what to do, and more or less how to do it. But other problems like genetically engineering sheep to produce better wool, superficially no more complicated, will be as unrealizable in a decade as they are today, because we have very little idea of where to start. Only when some fresh discovery or technique gives a new angle on the problem will anyone be able to say, 'Ah, *that's* how we can do it', and then we can start the long grind to actually achieving a result. Failure to realize the difference between these three types of future project has already resulted in a boom of over-enthusiasm for animal genetic engineering, which will result in disillusionment and loss of interest when the expectations are not fulfilled. I hardly need a crystal ball to see this far ahead, as the same patterns have dogged applied science since the First World War. There will be as little reason for both enthusiasm and disillusionment as there was for the analogous fads for bacterial genetic engineering, and 'ecology' before that, and nuclear power before that (remember 'electricity too cheap to meter'?). But such waves are fuelled not by reason but by hopes and fears.

So let us also move beyond the realm of cautious prediction, and into the realm of hopes and fears. Where is genetic engin-eering taking us? Once we have overcome the immediate techni-cal problems that we have been discussing so far, what sort of world shall we see around us? The next two chapters give my own idiosyncratic glimpses into that future.

Chapter 12

Dreams . . .

We can never predict the future, as weather forecasters frequently demonstrate. To try to predict the maturity of genetic engineering, a science now barely out of its cradle, would be folly indeed. But we can follow a few of the more obvious developments in the facts and speculations of today and see where they lead us. This will not be to predict the future: the path of genetic engineering will be influenced by new discoveries which are inevitably unexpected, and by social pressures which are beyond the technical description of the subject. Who would have predicted thirty years ago that the most conspicuous product of the electronic revolution would not be robot housemaids, but video games and word processors? So the following is just a small corner of a canvas which we will not see in its entirety for some time.

We shall look at the same three groups of organisms that we have considered before: bacteria and yeasts, agricultural plants and animals, and man.

In principle a bacterium can perform any chemical change that ICI or Dow Chemicals are already performing, but in a warm soup instead of in steel pressure vessels, and using sugar or sunlight as an energy source instead of oil or electricity. For many processes this would be cheaper in raw materials or power than the present-day method, and so would be very attractive for the chemical giants if the right bacterium could be made. So far no one has made a major chemical product more cheaply by using a genetically engineered bacterium or yeast, but this is due to insufficient experience, not to lack of potential.

Industry might start with bacteria to destroy its waste products before it allows them to work on saleable goods. As we mentioned in Chapter 8, some pollution disposal problems can be addressed by bacterial treatments already, but only a few chemicals are involved – ones which bacteria attack naturally, but which we find hard to remove. Some inventive genetic engineering will be needed to dispose of more general chemical wastes.

We might consider disposing of the pesticides which have been the focus of conservationsts' wrath for many years, as they can persist in the soil or on plants long after they have killed the pests at which they were aimed. Even if we stopped using them today, the problem of pesticides in the soil would still be with us in the year 2000. And pesticides are not the only long-lasting chemicals which can contaminate soils; many contaminants from chemical waste tips have spilled into the headlines, as at 'Love Canal' in the United States, where an entire housing estate near Niagara was dangerously contaminated by chemicals from the waste tip on which the houses were built. There is no way to get rid of such chemicals once they have seeped into the ground. There is no test-tube big enough to hold millions of tonnes of contaminated soil. In the past the only solution was to dig up the contaminated earth and bury it somewhere else. Genetic engineering offers an alternative: the creation of a bacterium that could break down the poisonous molecules in the soil. A bacterium which destroyed such chemicals on site could be sprayed on to the soil and left to work on its own, seeking out the poisonous molecules where they lay and destroying them, and simply producing more bacteria to add to the soil's natural humus content as a product.

This naturally leads us to think about using the same bacteria on the chemicals before they arrive at the dump site, so rendering the waste safe before it is dumped. This brings us back to bacteria being used to destroy waste. In principle most waste that is presently tipped, buried or burned could be disposed of by bacteria. 'All' we need to do is to genetically engineer them to make the right enzymes that will break open the molecules which we do not want and turn them into something harmless. Even better, we could try to turn them into something useful, for example we could turn waste paper into oil or alcohol.

However, the idea of using bacteria to dispose of waste is not really novel, even if the techniques we can employ to realize such an idea are new. More innovative is the use of bacteria in manufacture. We have mentioned that some uses of bacteria to make industrial chemicals are attracting a lot of interest, although not much commercial success. In addition to the well-established research projects to make propylene oxide (see page 96) and sugar products (see page 110), other uses of bacteria are more speculative but are still being actively researched by chemical companies. These include the use of bacteria to make oil from waste materials, alcohol from wood pulp, and several novel plastics. Bacteria could also be used to extract metals from ores, or even from mine tailings. They are already responsible for some metal mining processes: the nodules of manganese that litter the deep-sea floor are the product of bacterial action on seawater, and a process which extracts copper from poor-quality ores uses a combination of bacterial action and acids to release the metal.

Here bacteria are not making a process possible – it is already possible to extract manganese, or gold or uranium, from seawater. Unfortunately, it is far too expensive to be useful. Bacteria can make these processes practical by actively seeking out the few atoms of metal in seawater or in a poor ore, as a lion would actively seek out a deer in fifty miles of savannah. A lion would take no more than a day to catch that deer: consider the number of traps we would have to lay in fifty square miles to catch only one deer in the same time, and you can see why a living creature, moving actively after its prey, is much more effective at retrieving rare objects than a passive device, whether those objects are other animals or metal atoms.

The same capabilities open up a whole range of chemicals to bacterial synthesis. Complex chemicals are usually difficult to make in the laboratory, as the number of small changes which turn the starting material into the finished product are not very efficient. If each step is 90 per cent efficient (which is a better score than in many cases), then after ten steps only one third of the starting material is left. But bacteria can be much more efficient than this, converting 99.99 per cent of a starting material into a product. This is because their enzymes are very precise, and only change *one* starting material into *one* product. This applies to relatively simple molecules like food flavourings or colourings (see Chapter 8) as well as to much more complex ones. The most dramatic example is the duplication of DNA itself. A chemist can make DNA in the test-tube with an efficiency for each base of about 99 per cent. This means that if he tries to add one gram of a base on to the end of his DNA chain, .99 of that gram will actually get there. This is astonishingly good by the standards of normal chemistry, but it is still not good enough for a bacterium. By the time the chemist has made 100 bases he has lost 55 per cent of his starting material, and to get to base number 4,200,000 (the number of bases in the DNA of *E. coli*) he would have to start off with a truly staggering amount of starting material (in tonnes, 1,000,000,000 . . . and so on for another 3,711 zeros!) to get just one molecule of the right DNA out at the end. This is not a practical proposition for the chemist, but *E. coli* can manage to make such a vast molecule once every thirty minutes. This extraordinary ability could be turned to making less exotic materials like rare food flavourings or dyes, or novel materials to make micro-circuits or fabrics. Today such materials are known, but are so difficult to make that they are only laboratory curiosities. With bacterial help they could be major products of the future.

Introducing such exotic materials or using bacteria to produce more mundane ones would alter the whole environment, not just our chemical industries. This could be turned around to our advantage – bacteria could be used to alter the environment deliberately. If we allow our imagination free rein (and that is what we must do, because this is far beyond our present capabilities) we could envisage bacteria which are designed to fertilize the Sahara or thaw out the tundra of Siberia. The former would be performed by bacteria which were avid at extracting water from the air (even in the Sahara the humidity rarely falls below 40 per cent) and which, when they die at the end of their life-cycle of only a few days, keep that water trapped in humus-like compounds in their cells. The tundra could be thawed by spreading any black compound over its permanently frozen soil. The reason why Arctic soil remains frozen even during the summer is not that the air is too cold to thaw it out but that the soil itself reflects all the heat, as it is covered with a layer of snow and frost which acts as a mirror to the sun's rays. If we covered that mirror with some soot the heat would be absorbed into the ground, which would gradually thaw out and become amenable to farming. Of course, covering Siberia with soot is not really practical; but it might be possible to spray it with a bacterium that produced a black pigment as it grew. Bacteria can survive such temperatures – some can survive frozen into Arctic lakes at eight degrees below freezing – and bacteria produce the dark pigments in peat.

Would we want to do that – to unleash genetically engineered bacteria to alter our world? Again, once we have solved the technical problems the methods of genetic engineering become almost unlimited in their potential, and the limits on what we do are set by the larger society.

The options for the future genetic engineering of yeast are as varied as those for *E. coli*, because both these micro-organisms can be used for a wide variety of industrial processes. Which of the two we choose to employ depends on relatively minor technical points (like whether post-translational modification is important or not). So in our rather longer perspective, yeasts are likely to be used for the same sorts of project as bacteria would be.

With the genetic engineering of yeasts and bacteria now in the early stages of industrial respectability, scientists and professional prognosticators have been turning their attention to plants and animals. Here the field is wide open – as yet we know only a few techniques for successfully altering plants and no reliable ones for altering animals. The course we can plot for these two areas is correspondingly more vague.

The most obvious projects parallel those which we have already performed for bacteria: placing single genes and their promoters into plants to make a protein, maybe an enzyme, that we need. The simplest example is to put a protein into a crop plant simply to improve its food value. This could give the plant the nutritional value of beef, although it may not taste too good. With more complex techniques, we can go further. Wheat cannot grow in the far north or south because it is adapted to the cycle of winter and summer seen in temperate latitudes such as that of England. But there is little reason why wheat should not grow in Siberia, with careful farming. It will not do so without help, as the summer days are too long and the winter ones too short, and the wheat plant's own calendar which tells it when to sprout and when to fruit in more southerly climes would be confused.

Only a small number of chemicals are involved in controlling this calendar. These act as hormones in the plant, and their production is probably controlled by only a few genes. By altering these genes, scientists could override the wheat plant's own calendar preferences and make it grow anywhere where the ground was soft enough to plough. Large areas of the earth now considered incapable of sustaining agriculture would thus be available for food production. This project would need a complete understanding of how to make a gene work at a precise time in a precise cell in the plant, and thus it is quite unattainable today. However, it is easy to see how present-day research could lead to the knowledge necessary to perform such an operation.

A project like this would not affect the overall shape of the plant, at least not directly. But engineering wheat so that it would grow in the tropics would necessitate altering the enzymes the plant produced, the cells in which they were produced and the place in the plant where those cells lay, this last causing some small changes in shape. Wheat does not grow as successfully in the tropics as maize because it lacks a complicated mechanism that makes the best use of brilliant sunlight. The enzymes of this mechanism are present in special cells in maize, and it is therefore a staple food in many tropical countries. Just putting the same enzymes into wheat would be ineffective; the enzymes have to be near each other and the parts of the plant that use sunlight, directing molecules from the outside of a leaf into its core. To put this system of enzymes into wheat we would have to insert not merely the genes for the enzymes but also genes to tell the cells in which those enzymes reside where to grow in the leaf. This would be a much more complex proposition, although it is one we might want to undertake because wheat flour is of much better nutritional quality than maize flour. (An easier solution would be to improve the quality of maize flour by engineering the *maize* plant to make more protein, as we suggested in the last chapter.)

Similar problems may face scientists trying to engineer plants in order to allow them to 'fix' nitrogen. As we mentioned in the last chapter, giving plants the genes for enzymes which can 'fix' nitrogen from the air into nitrates would be useless because the 'fixing' enzymes need a cluster of other proteins around them before they can work. Making this happen is beyond our present technology, but we are already aware of the things we need to know to make it possible, which is the first step. It should become possible to dispense with the bacteria which normally perform the 'fixing' of nitrogen and produce a plant which can do it entirely on its own. This would require us to engineer a plant which was able to form a small pocket of proteins inside some of its cells where 'fixing' enzymes would reside: this would be difficult, but the effort could be rewarded by a much more efficient plant. (Curiously, one of the proteins involved is a sort of haemoglobin!)

These alterations are quite small ones, involving the addition of a few dozen protein molecules to a small number of cells. As we mentioned in the last chapter, plants will probably be able to cope with such manipulation without special precautions to deal with any side-effects. However, larger changes will need to be made with more care in plants, and even small changes are liable to produce adverse effects in animals. Will we be able to carry out genetic engineering to perform dramatic alterations in shape, alterations which are visible to the naked eye rather than just through a microscope?

To do this, we must understand the broad implications of our engineering – not just what our altered genes will do *in* the cell, but what they will do *to* the cell. Thus we must understand the way in which genes control the shape of cells, and how those genes themselves are controlled. To say that genes are responsible for overall cell shape is one thing, but to say how that responsibility is discharged is quite another. How are we going to make sure that out of the 6,600,000,000 bases of DNA in a human cell there is not a gene whose action depends on something we are disrupting? At the moment we do not know; we only know the order of a few hundred thousand bases of human DNA anyway. Even this is too complicated to understand without the help of a computer, so the examination of all the genes in a cell to see how we can alter that cell without harming it will require powerful supercomputers as well as advanced genetic engineering.

The construction of a cell may make such engineering impossible in some cases. There are alterations one can perform quite easily on an automobile – putting on new headlights, tuning up the engine and so on – but others which are impossible simply by altering the original. An automobile cannot be turned into a satisfactory boat without effectively rebuilding it from scratch. Similarly, we might be able to alter animal genes up to a certain point but be unable to take them beyond that limit without building an entirely new organism – a project so complicated that mapping every grain of sand on the seashore would be an afternoon's relaxation in comparison. However, let us assume that this is not so, and that with sufficient knowledge we can alter any animal in any way we want. What shall we do with such powers?

There are some obvious possibilities. We could engineer plants to grow faster and to put more energy into their commercially useful parts. This would require more than just engineering those plant hormones we mentioned earlier. The amount of energy a plant puts into growing at a particular time of year depends on the ratio of leaves to stem and stem to fruit. Foresters would be particularly pleased with such a project, as the thirty years it takes to raise the fastest-growing tree might be cut to less than ten. Indeed, one whimsical suggestion has been made that the fast-growing trees could even fell themselves too. The tree would suck so much nutrition out of the soil around its roots that the earth would become very sandy and weak after a few years' growth; the roots would be engineered to be of such a size that when the tree reached a certain height the sandy soil would give way and it would fall over of its own accord.

However, such projects are not actually needed at the moment, as the world can produce quite enough food and wood to satisfy its needs. People starve largely because they cannot get the food that exists, not because it is insufficient. Lack of communication is more crucial than lack of calories.

The genetic engineering of plants could alter the communications industry too. The target here is the wall around plant cells. Outside their plasma membrane, plants form a thick wall made of many complex materials of which the most common is cellulose, which, when processed, makes paper. The architecture of the cell wall is amazingly complicated and on a tiny scale, with molecules laid down in complex networks visible only with the most powerful microscope. It has taken scientists in the micro-electronics industry decades to engineer 'chips' with such detailed patterns that a small computer can be placed in a piece of plastic only a couple of centimetres long. But plants make cell walls far more complex than this and on a far smaller scale. Electronic scientists would like to miniaturize their creations even further to reduce the electricity they consume and to increase their speed. One way of doing this would be to engineer plants so that, instead of cellulose, they laid down their fantastic wall designs in silicon to an electronics engineer's specifications. This would open up a strange new science somewhere between physics and biology, with the physicist designing circuits for a genetic engineer to build. Biologists would become computer programers, running computers which flower once a year. Simple plants would probably be easier to manipulate than complex ones like oaks or daffodils. Man might lose his title as the planet's brightest thinker to kelp!

By this distant time we shall certainly be able to engineer novel proteins to order (something we cannot even start to do now). Perhaps we shall be able to dispose of the plant entirely and reduce our electronic components to the smallest size possible by making them out of a single protein molecule. A protein molecule could be the equivalent of an entire micro-chip, and the proteins in a bacterium would be as powerful a computer as one which takes up an entire room today. Of course, you would need a very small keyboard to program it . . .

Curiously, those professional pedlars of fantastic futures, the science-fiction writers, have largely ignored the plant world, concentrating instead on the engineering of animals.

The plausible options seem rather more limited for animals than for plants or bacteria. As we mentioned, there is little use in engineering an animal to produce a single protein – for food or medicine, or for industrial use – if we can get a plant, yeast or bacterium to do it for us, and almost invariably we can. More useful would be to alter an animal's shape or the functioning of one of its organs, and this runs into the problems of the compatibility of our genetic engineering with the existing genes in the target animal. But let us assume here that this has been overcome. What could we do?

Some obvious options concern the 'improvement' of animals for food, for example by altering the fats in cows' milk to make the milk richer, or the muscle structure of bullocks to make their meat more tender. This would be a much more complex task than simply making the animals bigger, and would be more likely to be of lasting use to the farmer. Once established, such agricultural genetic engineering could have several apparently retrograde effects. Improved draught-horses could once again become economic alternatives to tractors for certain terrains, especially in an age of rising oil prices. Cats would be used to kill rats in preference to Warfarin. Dogs are already used by police and army to detect drugs and explosives: their role could be extended to search for leaky gas mains or discover ores of rare metals.

Some of these possibilities rely on the ability to alter mental attributes as well as physical ones. While a genetically engineered wheat plant only needs to have the right enzymes, a genetically engineered draught-horse must be easy to train and moderately intelligent, as well as having the right amount of muscle. Relatively simple changes can have drastic effects on human psychology despite their lack of sophistication. Patients with large areas of their brain damaged as a result of injury or disease can show characteristic changes in their mental abilities and character which depend in part on the bit of the brain that is damaged. Similarly, genetic changes are known which affect the sufferer's character in precise ways. Children suffering from Down's Syndrome, for example, are usually quiet, biddable and happy, and they love music, although why an extra chromosome number 21 should cause these particular changes is quite mysterious. However, the one noticeable thing is that all such simple changes lead to a diminution of intelligence. While this is considered a handicap in humans, a farmer might be very pleased to be able to give a bull the bovine equivalent of Down's Syndrome, replacing the cantankerous animal he has today with a placid, happy one. And a happy animal is more likely to be a fat, healthy animal. So there could be considerable incentives for attempting to genetically engineer the mental attributes of farm animals, even if the simple beginnings of such engineering would almost certainly have the same effect as most simple alterations to the brain, that of reduction of intelligence.

How we could change an animal to increase its intelligence is a much harder question. It is easy to make an automobile go more slowly – puncture the tyres, pour water in the gas tank, tie a large rock to the back. But making it go faster is the task of a skilled mechanic. Similarly with the brain: several sorts of quite non-specific damage can reduce intelligence, but the alterations needed to increase it will be fiendishly complex. The increase of intelligence through genetic tinkering is not brought any nearer by the complete inability of students of intelligence to decide what intelligence actually is, let alone what causes it. This is one of the problems inherent in the attempt to decide how much the variation in people's intelligence is due to their having different genes, and how much to their being brought up in different environments. The participants in this debate cannot even agree what they are arguing about. We have a long way to go before we can even say what we really mean when we say that we want to make an animal more intelligent, let alone before we try to decide how to do it.

Nevertheless, making animals more intelligent would be so useful that it is likely to be a goal of genetic engineering until it is either achieved or definitely proved to be impossible. Projects such as increasing the intelligence of apes so that they are able to take over simple jobs from man (which was first suggested by

Arthur C. Clarke and popularized in the film, *Planet of the Apes*) are unlikely to be popular because to a certain extent automation has already done this, leaving millions of people jobless. More socially acceptable would be the engineering of simpler animals to take over tasks presently done rather inadequately by computers. Many animals are already superbly adapted to perform very complex tasks, such as an eagle's ability to see a mouse in a field from a hundred feet in the air. Small manipulations could be used to harness these abilities, so that we would see eagles surveying power lines for insulator wear, pigeons judging the ripeness of fruit on a conveyor belt and bats testing hi-fi systems.

The technical barriers here are huge. Not only must the control circuits of the genes be unravelled – itself a herculean task – but the operation of the brain must also be intimately understood before our changes can hope to do more than damage that fantastic mechanism. Apart from the moral problems involved, the practical difficulties of such tasks are likely to put us off for many years. Consider that today we have identified only a handful of the neurotransmitter proteins with which the brain probably teems (see Chapter 3). The majority remain unidentified. Each one will have specific target protein (unidentified) on a few cells (also unidentified). The cells link up in a way which is neither exactly laid out in a rigid plan nor totally at random – what guides them? As even the number of different types of cell in a mammalian brain is not known, it is clear that we have a long way to go before we can say why, when we put a vision of a bone into a dog's brain *here*, nerve impulses to wag the tail come out *there*.

This is hardly a problem to be worked out on the back of an envelope. The complexity of the computer which can work out what the brain of a mammal is doing will necessarily be about the same as the complexity of that brain. So the study of artificial intelligence and natural mind will converge, each using more and more of the techniques of genetic engineering to generate the materials it needs to reach the common goal. Some strange times await us. One day we will be able to program a computer with a model of a dog's brain, and see exactly why the dog does what it does. The computer will be made not of silicon chips but of tailor-made molecules, probably the product of a genetic engineering which is almost unimaginable today. The result may look more like a sunflower than a metal box. Will we have created a sunflower that thinks it is a dog?

When that day comes, we will no doubt be able to ask it that question. Until then, let us return from the borders of fantasy to the last of our categories of genetic engineering. Talk of intelligence focuses our minds on what we believe to be the most intelligent animal on earth. Can we engineer our own genes?

On a purely technical level, the answer to that must be 'Yes'. The experiments already performed on mice and rabbits could have been done on human embryos. But of course they were not, because technical limitations are not at issue here. The path to human genetic engineering is one which, more than any other aspect of the technology, is fraught with legal and ethical problems – which we have deliberately skirted around in this book because our aim was to find out what was technically feasible, not what society might want. Discovering the latter will be an increasing preoccupation for lawyers, scientists and every other member of society now that the technical problems of genetically engineering human beings are being solved. If we go beyond what is technically feasible today, as we have earlier in this chapter, then the removal of the technical barriers simply lays the social ones open to clearer view.

Their removal also opens up a wide field of potential gains. The cure of Lesch-Nyhan Syndrome and AMA deficiency (see page 161) are just the start of the genetic cure of a wide range of diseases which doctors can only diagnose today. How many diseases have a major genetic cause is unclear but, as we mentioned in Chapter 5, diabetes, heart disease and cancer may be among them. In Chapter 5 too we were discussing the possibility that recombinant DNA techniques could be used to diagnose a predisposition to one of these conditions – a possibility within sight of today's technology. With our eyes on the unprovable future, we might extrapolate from this to suggest a cure for those weaknesses using some future development of that same technology. These improvements in the human condition would surely be welcome.

But once we move into the realm of 'improvement', the practicality and desirability of the ideas becomes less clear. We can envisage genetically engineered giants, strong men, people adapted to life in the sea or at the Poles. But such plans tell us nothing about whether our creations would thank us for tinkering with their genes, or whether they would be anything other than parodies of ourselves. The experience of many countries shows that a minority created or imported to perform specific labouring tasks can have a very unhappy history. So where do we draw the line between alleviating human disease and producing a machine with human ancestry?

This may seem obvious at first sight, but often appears less so on reflection. The control of the oncogenes, whose aberrant action may be the cause of cancer, would be universally applauded, but, as we mentioned at the end of Chapter 4, to achieve this would take us a long way towards abolishing many of the changes that occur in old age. Ageing is a process during which the body's cells lose the ability to divide in order to generate new tissue, replacing that which has worn out. But cancer cells never age in this way. Control of cancer could result in giving the body the ability to regenerate itself almost indefinitely. Few young people want to die, but the old sometimes feel grateful that their life is unlikely to go on for ever. The upheaval we would experience in becoming a society in which only rare diseases or accidents ever killed anyone would be extraordinary and would a 200-year-old man really be as flexible and original as a 20-year-old, even if their bodies were indistinguishable? Would an extra 130 years make us become even more set in our ways and even less lovable than old people may sometimes become? Even a cure for cancer could carry a sting in the tail.

A new ability is usually thought of as an asset to those who acquire it, and so we could direct our genetic engineering towards giving our children new capabilities which no one enjoys today. We could increase their range of senses, making them able to see ultra-violet light, for example (in principle, a fairly simple change in the proteins which make up the lens and cornea of the eye could accomplish this feat). Or we could increase the number of nerves to the muscles of the arm and hand, making them extraordinarily dextrous, or build up the mass of muscle in the limbs creating supermen (or over-muscled thugs, depending on your point of view).

And maybe, by genetically engineering their brains, we could make them happy with the result. But is this something we should do?

The most ambitious project, one which is often held up as an example of what 'they' will try one day but which, as we have mentioned, is the most difficult project we can imagine, is to increase human intelligence. That, perhaps, is the least inviting option, and not only because evolution has tended to be harsh with animals that over-specialize in any one attribute and it is unlikely that huge brains will be an exception. The most worrying thing about such a plan could be the gleam in the eyes of the creators of these geniuses as they contemplated a new order ruled by their supermen offspring. Such gleams have often presaged the darker episodes of human history.

Indeed, the future of genetic engineering seems so fraught with perils when we consider that we ourselves might be the subjects that we may wonder if all the roseate optimism of the earlier part of this book is really justified. I think that it is. The moral problems of the genetic manipulation of human intelligence are esoteric ones indeed for the science of today, especially as we have still not solved the legal problems which arose out of the use of bacteria in genetic engineering. But they are one extreme of a line of thought that sees a darker side to all genetic engineering, even the manipulations which we consider routine today. We would not have completed our survey of what is possible in the field of genetic engineering if we did not include some discussion of what could possibly go wrong.

Chapter 13
. . . and Nightmares

Our genetically engineered future is not entirely Utopian. Nightmares of catastrophe lurk among the dreams, nightmares which opponents of genetic engineering are as eager to exploit as its supporters are to ignore. Many of the problems are highly speculative. In the spirit of the rest of this book, we will try to survey some of them realistically and separate the horrors from the hype.

As we mentioned in the last chapter, there are many worries about the possibility of applying recombinant DNA technology to people. But these are hardly worries which we can address today, as we have no idea what technology we would use for such genetic engineering and hence no idea what the potential results could be. Our cares are wasted if they are directed at this never-never-land of speculation, especially if existing technology has its own, realistic, problems attached. And it has.

There are three general areas of fairly immediate concern. The first we mentioned in Chapter 7: it is the problem of the safety of present genetic engineering techniques. The second is the possibility of biological warfare, and the third concerns the social implications of this new technology.

Although the genetic engineering which we perform today may seem crude and unadventurous to our descendants, nevertheless it causes worries about how it might go wrong which are very real and quite justified. These concerns have led scientists to instigate the safety measures which we outlined in Chapter 7, and which have subsequently been backed up by various legal measures in different countries. The most obvious safety meas-

ure is the physical containment of the genetic engineering laboratory – locking potentially dangerous experiments in bacteria-
proof rooms so that their products cannot escape to menace the
population. The same technology protects hospital workers from
the infectious diseases they deal with, and so it is well tested.
This is still an important aspect of the safety precautions taken
today. The use of genetically engineered *Pseudomonas syringae*
on crops (see page 115) was blocked not because the bacterium
was dangerous but because the experiment was a step outside
the laboratory.

A more subtle precaution is 'biological containment'. Here,
biological manipulations are employed to ensure that the organisms used for genetic engineering cannot survive outside the
laboratory even if they do escape: an invisible, but none the less
impenetrable, 'wall' is built around the experiment to complement the physical one. In particular, bacteria and yeasts which
can grow only in laboratory conditions and which are too feeble
to survive in the outside world can be used for experiments that
might be hazardous. Similarly, vector molecules can be engineered which carry mutations incompatible with the workings of
normal bacteria. Thus they will only be duplicated in the special
laboratory types of bacteria for which they were designed, and
cannot spread from them into other bacteria in the outside world.

But genetic engineers have discovered that many of these precautions are unnecessary. Genetically engineered bacteria are
feeble entities even without further crippling by the experimenter, and tend to die off, or to get rid of their new genes as
fast as they can. The same is true of higher organisms, too; the
genetically engineered version, although it might be an 'improvement' on the original from our point of view, is usually less
suited to survival in the outside world than the original. This
can be very irritating for scientists as it means that they have to
protect and cosset their creations to prevent them from being
overwhelmed, but it is rather reassuring for everyone else.

Thus the only time a genetic engineering experiment might
cause any real damage would be if a bacterium should pass its
new, engineered genes on to another bacterium, but without the
vector DNA which would have those lethal mutations in it, or if

the bacterium could do enough damage in the short time it was in the outside world before it died off. The first option is quite unlikely, especially as the 'wild' bacterium would end up by being burdened with the 'new' genes and so would be less able to survive than before, thus leading to *that* bacterium dying out. The second possibility is only likely to be a problem if the genetically engineered bacterium were producing a very dangerous molecule, for example a poison like botulinus toxin, the protein which kills the victims of botulism. Less than a millionth of a gram of botulinus toxin is enough to kill an average adult if it is injected (although far more would have to be made by a bacterium in a sewer before human disease appeared). However, the possibility remains that a bacterium in which we had put the gene for botulinus toxin would do some damage if released into the outside world; consequently the committees that regulate the applications of genetic engineering are very cautious about allowing anyone to perform experiments as potentially risky as this, and only permit the cloning of genes for very dangerous proteins in the most carefully sealed laboratories and in the most feeble of bacterial strains.

This 'escape of the demon bacterium' scenario is only the most crude possibility. Restrictions on what experiments are performed, coupled with careful tests of genetically engineered products before they are allowed into the outside world, will remove nearly all risk from this quarter. Such testing is obviously necessary, and there is a strong commercial incentive for it. No farmer is going to allow his crops to be sprayed with a genetically engineered bacterium if there is even a remote chance that it is not safe, as that risk would stop people buying his product. Even the combined might of the multinational chemical companies is being put under stress by the people who claim that the chemicals the companies produce have damaged their lives. Both the Vietnam veterans who were sprayed with the herbicide Agent Orange and the Indians whose dwellings clustered around the chemical plant at Bhopal believe that the chemical companies which manufactured the chemicals that blighted their lives have a duty to pay them huge compensation for loss of health, and some courts are supporting them. The lesson will not easily be forgotten – genetic engineering will have to be capable of being shown to be harmless in court before major companies will use it in the outside world.

But merely being safe in laboratory tests may not be enough. We could genetically engineer a bacterium to flourish on, say, dioxin, the poisonous chemical which contaminated the Italian town of Seveso and which Vietnam veterans claim was in Agent Orange. Such a bacterium would be a real boon for cleaning up waste tips and contaminated soil. Once the dioxin had been destroyed, the bacterium would have nothing left to eat and so would stop growing. We hope so, anyway, but although it may work that way in the laboratory, no one can really test it on a 10,000-tonne rubbish tip there. So the final trial will have to be a series of field tests, each getting nearer the actual conditions in which it is to be used; the field trial of the *Pseudomonas syringae* which we mentioned in Chapter 8 was planned to be part of such a series. Bacteria are complex organisms, and might defy the scientist's expectations of what they are going to do. We can be sure that they would not be poisonous, but they could be a nuisance. Even if they just lay around in heaps, bacteria are less than exciting additions to the landscape, and they usually smell bad too.

Such problems seem worrying until we pause to look at the precedents. Has such an explosion of bacterial growth actually occurred before? Yes, it has. Bacteria are used *today* to destroy sewage, and furthermore those bacteria are far healthier and stronger than are our genetically engineered ones. There are already bacteria which eat nylon, but the underwear industry does not lose much sleep over them. Occasionally these organisms, some of which can produce quite virulent poisons, do grow out of control: the drains block up, the towels start to smell, the steak goes bad. The result is invariably smelly and unpleasant, but is well within the capabilities of quite simple technology to clean up. There is no reason to suppose that a genetically engineered bacterium would even be this hard to dispose of.

Microbiologists, scientists who study micro-organisms like bacteria, came to a similar conclusion some time ago. They routinely handle dangerous bacteria and viruses, but they know that their techniques are adequate to cope with the risk and to protect them, and thereby protect everyone else as well. Indeed they were rather annoyed when all the fuss arose over the potential dangers of genetic engineering. Of course there are potential dangers, they said, but no worse than we have successfully coped with for decades.

In short, it is very unlikely that the genetic engineering which is being carried out today or is likely to be carried out in the foreseeable future will add any significant problems to those which the bacterial world already cause us. But a grimmer spectre is raised by the related possibility that a harmful bacterium could be released not by accident but by design, as a weapon of biological warfare.

In the 1960s, when a biological warfare program was part of every self-respecting nation's defence budget, scientists pointed out just how useless a biological weapon would be if they ever succeeded in making one. The most advanced biological weapons then available, mostly either versions of highly infectious viruses which affected the blood or of anthrax bacteria, had two drawbacks: they did not kill the enemy and they did kill you. They were dangerous, certainly, but never lethal enough to knock out more than 20 to 30 per cent of even an unprepared enemy, compared with the much greater efficiency of atomic bombs, bullets or clubs. They also did not act quickly enough. Even the fastest-acting diseases take more than a day to begin to affect their victims, and in modern warfare entire countries can be laid waste in a day. So the weapons would be pretty useless against any modern army, although they might be quite efficient in killing citizens of Third World countries which lack Western medical facilities. And after the battle the bacteria and their spores would remain on the battlefield for weeks, would drift in the wind and would end up by causing as many casualties among the side that launched them as among the enemy, and many more among the civilians who had to clean up the mess afterwards.

Added to this was the difficulty of producing weapons that did not 'leak', and the problem that is very difficult to find any disease-causing organism that is strong enough to withstand being fired from a gun or exploded out of a bomb – out of 100 of even the toughest bacteria, less than one would be likely to survive delivery to the target. Thus biological weapons would be the opposite of modern nerve gases. 'Binary' nerve gases are mixes of relatively safe materials which react after they have left the gun to form a deadly poison; biological weapons would be lethal when they were being made, but largely harmless by the time they reached their target.

Given these objections, it is not surprising that those people who are for ever searching for new weapons gave up on biological warfare by the end of the 1960s.

The same problems apply today. A few bacteria and their poisons have been studied in much more detail so that we have a better idea how they work, but this has not encouraged the arms merchants. Some exotic uses of bacterial poisons have hit the headlines, as when the Hungarian defector Georgiou Markov was killed by a small pellet loaded with a bacterial toxin, fired from the tip of an umbrella. Delightful as the image of the British Army marching into battle with shrapnel-proof bowler hats, exploding briefcases and poisoned umbrellas may be, this does not seem a very realistic weapon for anyone other than a terrorist.

The molecular structure of some of these toxins is quite well known. Botulinus toxin, for example, is a protein. In principle we could genetically engineer an *E. coli* to make lots of botulinus toxin, and then use that in a large-scale terrorist campaign, dropping a few grams in a barrel of beer, for example (how terrorists could do this without poisoning themselves is another matter). But that sort of genetic engineering requires a great deal of technology, and any nation with that much expertise would want a far more useful weapon than one which could be used only to put a pub out of business.

Thus the barriers to the use of genetically engineered bacteria as biological weapons are probably greater than those which would militate against the use of 'wild' bacteria. The bacteria which cause disease are highly specialized for their role, and rely on very complex tricks to evade the defences of their victims. Anthrax or plague are not just *E. coli* with a poison added: they are carefully crafted invasion machines, and medical scientists still cannot explain why they cause disease while other bacteria live in or on us and cause no disease, and yet others would die within minutes of landing on our skins or in our stomachs. No one knows why an infection by rabies virus is 100 per cent fatal, while infection by rhinoviruses only gives you a cold. Until someone finds out, we are never going to be able to genetically engineer a 'new' disease even as effective as those that already exist, let alone one which is more lethal.

In short, World War Three seems most unlikely to be a biological war. Nature puts numerous obstacles in the way of those of her children who want to use her resources for killing each other.

Such are the facts. Every year or so, microbiologists and genetic engineers recite them in an almost ritualistic way to counter new suggestions by the more fanatical anti-science groups that we are on the brink of a biological Armageddon, or by political extremists that 'they' are using biological weapons against 'our' allies. The facts are also pointed out to the military planners every time they start to dream of weapons that will strike down the enemy but leave our side unaffected. There is no scientific disagreement about the inefficiency and undesirability of biological warfare. And yet the United States and the USSR are once again moving into this field with typical military myopia, each claiming that their growing research programs are purely defensive, a protection against the other's threat. The US Defense Department has openly advertised for scientists to work in its laboratories on aspects of biological warfare; the USSR is less candid, but has increasingly taken to declaring that work on recombinant DNA is secret (as nearly everything is secret in the USSR, it is hard to know what to make of this). In 1980, near Sverdlovsk in the north-western USSR, an 'explosion' was alleged to have occurred at a germ warfare production plant making anthrax bombs. The impact of the story was rather lost when the real source of the anthrax outbreak was revealed to be contaminated sheep: wool production is a major industry in the area, and sheep are the main carriers of anthrax. Clouds of 'yellow rain' were taken as evidence of the next supposed USSR biological weapon, a horrible spray which sent villagers in South-East Asia running for sanctuary to the CIA. Professor Matthew Meselson of Harvard showed rather convincingly that this dreaded weapon was actually bee faeces. The USSR regularly accuses the United States or Britain of researching, making or actually using biological weapons, as do left-wing groups in Europe. Again, proof is sadly lagging behind accusation.

This would be an excellent plot for a farcical film were it not for the scale on which these dramas are played. There is a real possibility that the military planners will come to believe their own propaganda and decide that biological warfare really *is* being waged against them – that a biological attack is actually a sensible option – and will load up a bomber with anthrax bacteria and spread it all over Poland or Nicaragua as a substitute for negotiation. Although anthrax is a useless weapon, it is a very dangerous disease. And if 'they' have used biological weapons, what is to stop 'us' from using anything else, maybe chemical or even nuclear weapons? Which, sadly, are much more effective. Thus the fear of biological warfare could generate as much destruction as the weapons themselves are meant to do, but cannot. The military threat from biological warfare is very small. But the threat from fear and misunderstanding of what biology can give the weapons-makers is real, and appears to be growing. It is ironical that this fear should be the most dangerous product of the recombinant DNA revolution, when that revolution has shown us many of the reasons why the fear is groundless in the first place.

The technical obstacles can be overcome, of course. Given enough knowledge, all dreams – even nightmares – become reasonable. In theory we could imagine a weapon which not only killed quickly and efficiently but also discriminated between friend and foe, as gas or nuclear fall-out never could. By turning the body's own defences against itself, a virus could, in principle, kill 70 per cent of orientals but only 25 per cent of occidentals, or be entirely specific for caucasians or negroes. Mercifully, such ideas are only at the science-fiction stage at the moment. However, it is the sort of science fiction which encourages professional doomsters to write headlines of the 'Science Dooms Globe!' type, because they are so much more eye-catching than the correct version, which would read: 'Science might doom globe in 100 years if we let the army get away with it'.

But the pace of research is frantic, and the desire of people to kill other people seems almost unlimited; so someone, somewhere, is bound to try out a biological weapon one day. Such a project could well abandon unreliable viruses or bacteria entirely, and switch instead to engineering a weapon based on a more discriminating organism like the mosquito. Indeed, the biological weapon would be one with enough intelligence to be 'programmed' with a complex battle plan and then to modify the plan if it went wrong, to be extremely tough and flexible, to be able to discriminate friend from foe even in the absence of obvious physical differences, and to be quite safe in between wars but to be so enthusiastic about killing the enemy in wartime that it would die itself for the chance. If that sounds to you more like a religious fanatic than a product of genetic engineering, it demonstrates the last reason why biological warfare is useless: we can do it already.

In this chapter we have looked at the possibility that something might go wrong with genetic engineering and decided that it was rather unlikely in the foreseeable future; we have also looked at the possibility of something going wrong on purpose, and concluded that it is not merely unlikely but pointless too. Is the genetically engineered future free of all clouds?

I think not, for two reasons. The first is that all such estimates of what will happen in the future make only predictions of chance. I know that it is very unlikely that I shall be struck by lightning tomorrow, but it is possible. We can never live in a totally safe world and every new technology adds to the uncertainty, although most remove more uncertainty than they create. There is an outside chance that, before the year 2000 recombinant DNA technology will cause an accident in which someone will die. It is also almost certain that that same technology will help to find treatments for AIDS and cancer, and diagnostic methods for many other diseases; as a result, thousands will live. This seems a worthwhile exchange. But the outside chance remains.

But if the one in a billion chance does not turn up, there is still a major problem, and one not addressed by any of the safety measures employed today. The worry is not that genetic engineering will fail, but that it will succeed.

197

This is an odd thing to say. After a bookful of description of everything that genetic engineering can do, or will be able to do, to improve our lives, how can I turn round and say that that very success is its biggest failing? In itself it is not, of course. It is a triumph of the application of the greatest scientific adventure of our age, the discovery of our own biological nature. But for all its revolutionary scientific and technical aspects genetic engineering is profoundly conservative, and is today being used to freeze many aspects of our society which need to change and which recombinant DNA technology *could* change. This not only stores up trouble for the future, but wastes the magnificent opportunity with which recombinant DNA presents us.

Let me take three examples. The production of human insulin using recombinant DNA technology was undoubtedly a great achievement, and has benefited millions of diabetics worldwide. Yet it will not reduce the number of diabetics in our culture even by one. Since the Second World War that number has grown systematically – in 1984 it was increasing by 6 per cent per annum – and the supply of insulin has had to grow with it. Diabetics can lead an outwardly quite normal life, but they need to keep their blood topped up with the vital hormone and are always on the lookout for the first signs of secondary problems of the disease, such as dying nerves in the limbs, or blindness. They seldom live as long as their unaffected brothers and sisters. Human insulin is a valuable addition to the treatment of the disease, but it is not a *cure* and even a cure would be second best to complete prevention. The true cure would be to remove the cause, which many doctors feel is to be found in a lifetime of eating high-calorie, high-sugar foods, too much stress and too little exercise. (I don't want to sound like a killjoy. Even diabetics must eat *some* sugar. But two pints of ice-cream before bed each night, which is what one US nutritionist said he ate before his reformation, is excessive.)

But not only does genetically engineered human insulin fail to address this problem, it positively aggravates it. It gives the impression that a miracle cure for diabetes exists (indeed, some tabloid newspapers announced the commercial sale of human insulin as a 'cure' for diabetes), and so there is no need to prevent the disease in the first place. There are no miracle cures in genetic engineering, or in any branch of medicine, only solutions to specific problems. As a solution to the problem of how to give a diabetic as near normal a life as possible, human insulin is excellent, and the scientists who produced it have performed a great service to science and humanity. But as a solution to the problem of how to stop people contracting diabetes, it is a washout.

The same difficulty applies even more strongly to other areas of recombinant DNA technology. The *Pseudomonas syringae* 'ice-free' bacteria could be an example of a neat solution to a rather unimpressive problem; another was announced last year by two genetic engineering companies. Calgene, a research company in California, and the agrochemical giant Monsanto were both working on ways of making crop plants resistant to the herbicide glyphosate. Glyphosate kills all green plants by blocking the action of a vital enzyme, so it is a wonderful weedkiller. Unfortunately, it is a wonderful cropkiller too. Calgene and Monsanto each have projects which aim at the same result but by different technical means. They want to alter the genes in a plant (ultimately wheat, although for technical reasons they are starting on tobacco and petunias respectively) so that the plant itself produces a lot more of the enzyme attacked by glyphosate, thus overcoming the effects of the herbicide. Even though 95 per cent of the enzyme will be knocked out by glyphosate, there will have been twenty times as much of it as there was in the original strain of plant, so the plants will remain healthy. Thus these plants will be resistant to the chemical. If they were planted as a crop, the whole field could be sprayed with glyphosate and everything growing in it except the crop would die: the perfect weed-control system.

Or is it? It is certainly a brilliant answer to the problem of how to develop a chemical spray system to keep down weeds. Such spray-and-fertilizer intensive agriculture has made the United States the breadbasket of the world, consuming 750,000 tonnes of pesticides alone in 1982, and bringing a substantial number of farmers there to the verge of bankruptcy. Intensive farming is destroying the soil and water resources in the central USA, and even conservative biologists admit that in fifty years substantial parts of the wheatfields of the Mid-West will be eroded beyond use, and about a third of the artesian water now essential to the farming of this region will have dried up. (This will bring about a catastrophic increase in the cost of food produced on these farms. A genetic engineering 'fix' exists to cure this too – the use of bacteria to make bulk protein for foodstuffs as pioneered by ICI's Pruteen. But this will not help the farmers either, and will only last if natural gas products to feed the bacteria are cheap.) The long-term problem is not how to get the maximum amount of food out of a field by killing off all the competing weeds. This will maximize farm output in the short term, but will do nothing for the longer-term prospects. The real problem is how to release the farmer from a complete reliance on high-energy, high-tech, high-cost farming which is draining money out of the farms and life out of the soil. Will the genetic engineering of crops to be resistant to glyphosate address *that* problem? No; again, it will simply encourage a temporary high-tech 'fix' which will ulti-mately aggravate the underlying problems. Indeed, it is hard to imagine a genetic engineering project which would attack the real problem and which would also be commercially attractive to the chemical companies funding the research. The best example is Cetus's project to engineer pest-resistant crops, thus reducing the farmers' reliance on pesticides. It is difficult to imagine the chemical companies, which support much agricul-tural research but which also produce the pesticides, looking favourably on such a scheme. The evidence is that they will not. While Agrigenetics and Plant Genetic Systems are trying to genetically engineer a plant which is more resistant to insect pests, Monsanto, a large chemical company, is using the same technology to develop a new spray to kill those pests. As with the glyphosate-resistant plant project, this simply makes the farmer dependent on yet another spray.

I have singled out these two examples not because they are in any way atypical, or because they show evidence of a plot to use genetic engineering to bring down the West. What they do demonstrate is that the use of genetic engineering to attack short-term goals – even extremely laudable goals such as the alleviation of diabetes – may turn out to be less helpful in the longer term.

The desire for a quick 'fix' has overcome the very real need for long-term planning.

It is probable that the researchers think that their particular project is the greatest thing since fire, and this is all to the good because nothing else would carry them through the years of uncertainty and hard slog which comprise any research project. But in the case of some of the projects it is questionable who the result will benefit. The two examples cited above have an obvious beneficiary in the short term, but a less clear longer-term prospectus. Other projects do not seem to be responding even to a short-term need, but only to a short-term profit. The genetic engineering of human growth hormone may be such a project. The estimated market for human growth hormone is quite easy to calculate, as the medical use of the hormone is confined to the treatment of specific well-known diseases. This market is estimated to be worth about $20 million in the United States. This is small by medical standards, and barely warrants the research effort being put into making growth hormone by the use of recombinant DNA technology. However, a manager of Celltech, the British genetic engineering company, estimated that the actual market for growth hormone in the United States would be at least $100 million. Part of this sum may be explicable by the hormone's potential use for injection into cattle to make them grow faster (the same result as would be achieved by giving the cattle the gene for growth hormone). However, most must be for human consumption. So where is the remaining $80 million-worth really going?

My only plausible answer is that there is a strong 'black market' demand for human growth hormone from fathers who want their sons to be 6' 6" baseball players, from daughters who have heard that no one in Hollywood is under six feet tall and so want to be 6' 2", for football coaches who want to give their team that extra edge. If you think it unlikely that pharmaceutical companies would stoop to sales of this sort, consider that the epidemic of abuse of such drugs as Quaaludes and amphetamines in the 1970s occurred partly because the pharmaceutical giants produced these drugs in huge excess over their real medical requirements. The steroid hormones used by some athletes were not produced in someone's kitchen: their synthesis requires the facilities of sophisticated chemical laboratories. Human growth hormone has a real advantage over steroids. It is normal human protein, so there is no way to prove that the athlete who is 7' 6" tall and has a bloodstream full of growth hormone had been injected with it a week before. She might just naturally have a lot of growth hormone.

Bear in mind that these examples are not hypothetical problems as all the others mentioned in this chapter have been, but real projects which seem to be leading recombinant DNA technology astray. What is lacking is foresight. When a doctor finds a chemical which might make a good drug, he does not just test whether it will kill his patients straightaway before he uses it. He also checks to see whether the drug actually cures rather than conceals the disease, and whether it has any adverse effects in the months and years which follow its administration. When you inject a major new technology into a complex society, surely the same criteria should apply.

Is it unreasonable to ask for this foresight? After all, if the companies which research and market the products of genetic engineering are not profitable in the short term they will go bankrupt and have no long term at all, so they can hardly choose to be altruistic. But I do not think that this is an unrealistic thing to ask.

We can look in the two quarters for enlightened self-interest in the long term. First, we should ask who pays for nearly all fundamental research in genetic engineering, and some applied research too. The answer in all Western countries (and also in communist ones, of course) is the government, which really means all of us. But the government does not need to be profitable in the short term. Indeed, the government should be there to protect its citizens' long-term interests. The government is in a position to insist that every dollar spent on producing insulin by recombinant DNA techniques in a government-funded laboratory is matched by a dollar spent on trying to prevent diabetes by improving public awareness of health problems. This principle could apply to a large proportion of fundamental research, as even such apparently non-governmental sources of finance as charities benefit indirectly from tax allowances. Even corporate R&D is tax-deductible.

But the companies themselves should also be looking to the long-term future. While a small company like Genentech may have to aim its research at short-term gains because without them it will collapse, the same is not true of the research efforts of such giants as ICI, Shell, Monsanto and General Electric. A recombinant DNA research budget is a small part of the funds of these vast corporations, and should be focused not on the short-term gain, which is the responsibility of the production side of the company, but on how to carry the company successfully into the twenty-first century. The way to do this does not include driving your customers into bankruptcy. Monsanto's insecticidal spray may bring short-term profits, but as farmers either divest themselves of their reliance on chemical agriculture or see their farms becoming unfarmable wastelands, Monsanto will find these profits turning to losses. In other fields, major corporations direct their most adventurous research to the distant future: oil companies sponsor research on wind and tidal power; computer giants research into computers which will not reach the production stage until the next century; aircraft manufacturers look into the production of spaceships for the time when they may be called upon to ferry holidaymakers to the moon. But genetic engineering does not seem to be favoured with such foresight. We are only looking to the short-term future, when it is in all our interests not to do so. With forward planning we could avoid using genetic engineering to exaggerate the short-sightedness and technological obsessions of our century and use it instead to create the basis for a genuinely novel industrial base. Unfortunately, those same failings militate against such a program being adopted. The tragedy, then, is not that genetic engineering will not work. It is that it will be used for short-term, trivial or even destructive purposes while its potential for good is thrown away, because making sure that the banks are there tomorrow will not put money in the bank today.

Chapter 14
Genetic Engineering for **Absolutely** *Everybody*

Once again, we have strayed from the technical aspects of genetic engineering into the sociology of science. The science of genetic engineering is not immediately concerned with such social issues. Like all such sciences, genetic engineering is a set of tools, a box into which men put resources, effort and enormous patience, and out of which comes the miracle that is the twentieth century.

Twenty years ago few had heard of genetic engineering. They were mostly science-fiction readers or eugenicists, each in their own way living in a dream world. Now genetic engineering is quoted on the Stock Exchange and captures some of the brightest minds of our age. In this book I have tried to give some idea of genetic engineering as a reality. In our pursuit of that reality we have delved into the world of molecular biology, the machinations of the molecules of life, and emerged again to see how malfunctions of our bodies and the engineering of bacteria could all be related to the central molecule of life, DNA. And because we can now alter that molecule, we can alter life itself to our design.

The power to alter genes at will, to step ahead of blind evolution and direct the change of organism to our own ends, is an almost magical one, and one we have hardly begun to use. Yet already commercial products are on sale employing this decade-old technology, and a growing band of projects follow in their footsteps. Beside them, scientists are opening up a treasure-house of information about how our genes operate, and how they fail. However else they are used, these results will be ploughed back into the technology of recombinant DNA to generate new methods, new results. For genetic engineering is not a business or a product, but a method. I hope that you now have an idea of where this method *can* go and where it is impotent, where a little knowledge will open a new door and where whole new sciences will have to be created before we can carry out our plans. The potential is as large as life itself, for genetic engineering is the alteration of the most fundamental parts of life.

Where *will* genetic engineering go next? I hope I have also made it clear that this is a problem which technicalities alone will not resolve. The decision about where we should go after insulin and 'giant mice' has not yet been made, because it is not a decision scientists can take. They can advise, and theirs is the technical expertise to execute the plan. But they can only say what is possible, and how long it will take to get there. We must provide the questions to be asked, the goals to be sought. Science is a tool, not a master. Genetic engineering is no exception.

Where we go from here is up to you.

Index